VIRTUAL CHEMLAB
GENERAL CHEMISTRY LABORATORIES
v.4.5

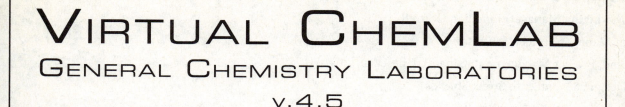

BRIAN F. WOODFIELD
MATTHEW C. ASPLUND
Brigham Young University

STEVEN HADERLIE
Springville High School

PEARSON

Boston Columbus Indianapolis New York San Francisco Upper Saddle River
Amsterdam Cape Town Dubai London Madrid Milan Munich Paris Montréal Toronto
Delhi Mexico City São Paulo Sydney Hong Kong Seoul Singapore Taipei Tokyo

Editor in Chief: Adam Jaworski

Senior Acquisitions Editor: Terry Haugen

Senior Marketing Manager: Jonathan Cottrell

Program Manager: Coleen Morrison

Team Lead, Project Management: Gina M. Cheselka

Cover Design: Seventeenth Street Studios

Operations Specialist: Christy Hall

Credits and acknowledgments borrowed from other sources and reproduced, with permission, in this textbook appear on the appropriate page within the text.

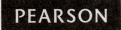

www.pearsonhighered.com

1 2 3 4 5 6 7 8 9 10—EBM—17 16 15 14 13

ISBN-10: 0-321-94327-9; ISBN-13: 978-0-321-94327-9

Table of Contents

Additional Assignments

Titrations

Gas Properties

Atomic Theory and Quantum Mechanics

vi

Overview

Introduction

Welcome to Virtual ChemLab, a set of realistic and sophisticated simulations covering general chemistry. In these laboratories, students are put into a virtual environment where they are free to make the choices and decisions that they would confront in an actual laboratory setting and, in turn, experience the resulting consequences. These laboratories include simulations of inorganic qualitative analysis, fundamental experiments in quantum chemistry, gas properties, titration experiments, and calorimetry. This overview outlines the general philosophy behind the virtual chemistry environment and briefly describes each of the laboratory simulations contained within Virtual ChemLab. Details on using the various laboratories are found elsewhere in the Virtual ChemLab Help.

Principles of Virtual Laboratories

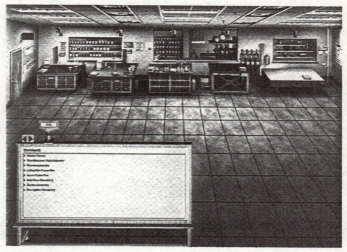

Disciplines such as chemistry, physics, and biology are laboratory sciences and as such require students to apply what they have learned in the classroom in a laboratory setting. Ideally, students should be placed in a situation where they can have access to level appropriate equipment and experiments to explore and discover the principles being taught in the classroom. In practice, however, such an environment is limited or does not exist because of cost, time, liability, and logistical constraints, and students are often placed in laboratory settings that encourage "cookbooking" and attention to procedures rather than the principles governing the experiments.

The challenge becomes balancing the competing demands of teaching students laboratory technique and the application of fundamental principles. While a real laboratory environment is ideal for teaching laboratory technique, the cost, time, liability, and logistical constraints limit or prevent exploration and open-ended assignments. On the other hand, virtual laboratory environments are generally not well suited for teaching laboratory technique since computers prevent "hands on" manipulations, but appropriately constructed virtual environments are ideal for exploration, access to expensive equipment, and allowing open-ended assignments and activities for students. The combination of real and virtual laboratories can provide the ideal environment for students to learn laboratory techniques and the application of principles in the design of experiments and the interpretation of experimental results.

The purpose of Virtual ChemLab is to provide a virtual laboratory where students are free to explore the various aspects of general chemistry and perform experiments in an open-ended environment. While simulations and virtual laboratories are often considered equivalent, Virtual ChemLab is more than a set of simulations since it creates a holistic and immersive virtual environment for students to perform open-ended inorganic qualitative analysis experiments, calorimetry, acid-base and potentiometric titrations, gas properties, and fundamental experiments in quantum chemistry. The primary focus in the virtual chemistry environment is to allow students to concentrate on performing experiments across the general chemistry discipline, collecting data easily, and then interpreting the data to understand fundamental chemical principles and theories. While some of the laboratories involve laboratory technique, that is not the primary focus and is best left to an actual laboratory setting. We want students to explore the classic experiments in general chemistry and as a result discover the underlying principles of chemistry by themselves.

1

Laboratory Overview

After starting Virtual ChemLab, students will be placed in a room with five laboratory benches that represent the five different general chemistry laboratories. By mousing over each of these laboratory benches, students can display the name of the selected laboratory, and clicking on the laboratory bench launches a projection of the selected laboratory where students can then perform experiments. While in the general chemistry laboratory, the full functionality of the simulation is available, and students are free to explore and perform experiments as directed by their instructors or by their own curiosity. Given below is a brief outline of what is available in each of the five general chemistry laboratories. Details on using the various laboratories are found elsewhere in the Virtual ChemLab Help.

Calorimetry. The calorimetry laboratory provides students with three different calorimeters that allow them to measure various thermodynamic processes including heats of combustion, heats of solution, heats of reaction, the heat capacity, and the heat of fusion of ice. The calorimeters provided in the simulations are a classic "coffee cup" calorimeter, a dewar flask (a better version of a coffee cup), and a bomb calorimeter. The calorimetric method used in each calorimeter is based on measuring the temperature change associated with the different thermodynamic processes. Students can choose from a wide selection of organic materials to measure the heats of combustion;

salts to measure the heats of solution; acids, bases, oxidants, and reductants for heats of reaction; metals and alloys for heat capacity measurements; and ice for a melting process. Temperature versus time data can be graphed during the measurements and saved to the electronic lab book for later analysis. Systematic and random errors in the mass and volume measurements have been included in the simulation by introducing buoyancy errors in the mass weighings, volumetric errors in the glassware, and characteristic systematic and random errors in the thermometer measurements.

Gases. The gases laboratory allows students to explore and better understand the behavior of ideal gases, real gases, and van der Waals gases (a model real gas). The gases laboratory contains four experiments each of which includes the four variables used to describe a gas: pressure (P), temperature (T), volume (V), and the number of moles (n). The four experiments differ by allowing one of these variables to be the dependent variable while the others are independent. The four experiments include (1) V as a function of P, T, and n using a balloon to reflect the volume changes; (2) P as a function of V, T, and n using a motor driven

piston; (3) T as a function of P, V, and n again using a motor driven piston; and (4) V as a function of P,

T, and n but this time using a frictionless, massless piston to reflect volume changes and using weights to apply pressure. The gases that can be used in these experiments include an ideal gas; a van der Waals gas whose parameters can be changed to represent any real gas; real gases including N_2, CO_2, CH_4, H_2O, NH_3, and He; and eight ideal gases with different molecular weights that can be added to the experiments to form gas mixtures.

Quantum. The purpose of the quantum laboratory is to allow students to explore and better understand the foundational experiments that led to the development of quantum mechanics. Because of the very sophisticated nature of most of these experiments, the quantum laboratory is the most "virtual" of the Virtual ChemLab laboratory simulations. In general, the laboratory consists of an optics table where a source, sample, modifier, and detector combination can be placed to perform different experiments. These devices are located in the stockroom and can be taken out of the stockroom and placed in various locations on the optics table. The emphasis

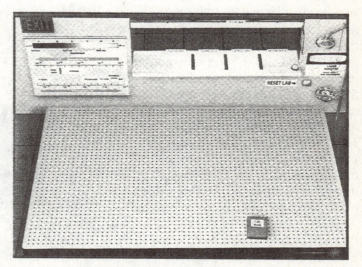

here is to teach students to probe a sample (e.g., a gas, metal foil, two-slit screen, etc.) with a source (e.g., a laser, electron gun, alpha-particle source, etc.) and detect the outcome with a specific detector (e.g., a phosphor screen, spectrometer, etc.). Heat, electric fields, or magnetic fields can also be applied to modify an aspect of the experiment. As in all the virtual laboratories, the focus is to allow students the ability to explore and discover, in a safe and level-appropriate setting, the concepts that are important in the development of atomic theory.

Titration. The virtual titration laboratory allows students to perform precise, quantitative titrations involving acid-base and electrochemical reactions. The available laboratory equipment consists of a 50 mL buret, 5, 10, and 25 mL pipets, graduated cylinders, beakers, a stir plate, a set of 8 acid-base indicators, a pH meter/voltmeter, a conductivity meter, and an analytical balance for weighing out solids. Acid-base titrations can be performed on any combination of mono-, di-, and tri-protic acids and mono-, di-, and tri-basic bases. The pH of these titrations can be monitored using a pH meter, an indicator, and a conductivity meter as a function of volume, and this data can be

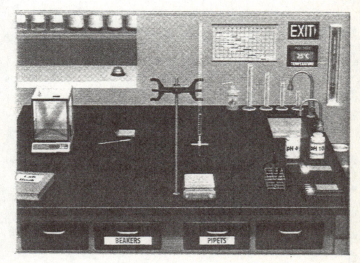

saved to an electronic lab book for later analysis. A smaller set of potentiometric titrations can also be performed. Systematic and random errors in the mass and volume measurements have been included in the simulation by introducing buoyancy errors in the mass weighings, volumetric errors in the glassware, and characteristic systematic and random errors in the pH/voltmeter and conductivity meter output. These errors can be ignored, which will produce results and errors typically found in high school or freshman-

3

level laboratory work, or the buoyancy and volumetric errors can be measured and included in the calculations to produce results better than 0.1% in accuracy and reproducibility.

Inorganic. The features of the inorganic simulation include 26 cations that can be added to test tubes in any combination, 11 reagents that can be added to the test tubes in any sequence and any number of times, necessary laboratory manipulations, a lab book for recording results and observations, and a stockroom for creating test tubes with known mixtures, generating practice unknowns, or retrieving instructor assigned unknowns. The simulation uses over 2500 actual pictures to show the results of reactions and over 220 videos to show the different flame tests. Cations that are available in the simulation include Ag^+, Al^{3+}, Ba^{2+}, Bi^{3+}, Ca^{2+}, Cd^{2+}, Co^{2+}, Cr^{3+}, Cu^{2+}, Fe^{2+}, Fe^{3+}, Hg^{2+}, Hg_2^{2+}, K^+, Mg^{2+}, Mn^{2+}, Na^+, NH_4^+, Ni^{2+}, Pb^{2+}, Sb^{3+}, Sn^{4+}, Sr^{2+}, Ti^{4+}, V^{4+}, and Zn^{2+}. Available reagents include HNO_3, H_2O_2, NH_3, Na_2S, $NaCl$, Na_2SO_4, $NaOH$, Na_2CO_3 and pH 4, pH 7, and pH 10 buffers. With 26 cations that can be combined in any order or combination and 11 reagents that can be added in any order, there are in excess of 10^{16} possible outcomes in the simulation.

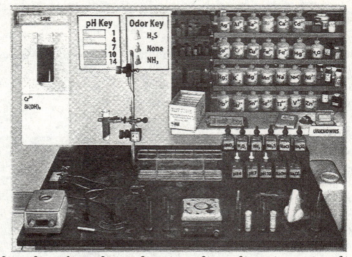

System Requirements

Minimum system requirements are as follows for v4.5:

Windows
Pentium III Processor (Pentium 4 recommended)

Microsoft® Windows® XP Home, Professional, or Tablet PC Edition with Service Pack 2 or 3, Windows Server® 2003, Windows Vista® Home Premium, Business, Ultimate, or Enterprise (including 64-bit editions) with Service Pack 1, Windows 7, or Windows 8

512 Mb RAM (1 Gb Recommended)

200 Mb of free disk space

Recommended minimum resolution 1024 x 768

Macintosh
Intel Core™ Duo or faster processor

Mac OS X v10.6, 10.7, or 10.8

512 Mb RAM (1 Gb Recommended)

200 Mb of free disk space

Recommended minimum resolution 1024 x 768

Installing *Virtual ChemLab*

Use the following steps to install Virtual ChemLab:

1. **Windows:**   Run the program "Setup Virtual ChemLab.exe" and then follow the prompts.

2. **Macintosh:**   Open the DMG file and drag the Virtual ChemLab application to the Applications folder.

Getting Started

Welcome to Virtual ChemLab, a set of realistic and sophisticated simulations covering general chemistry. In these laboratories, students are put into a virtual environment where they are free to make the choices and decisions that they would confront in an actual laboratory setting and, in turn, experience the resulting consequences. These laboratories include simulations of inorganic qualitative analysis, fundamental experiments in quantum chemistry, gas properties, titration experiments, and calorimetry. This section gives you a brief tutorial on how

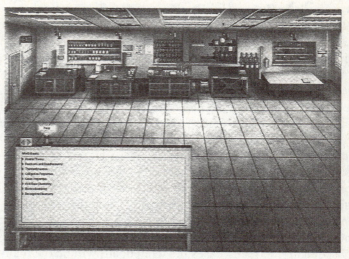

to access the different laboratory experiments and use many of the features found throughout the virtual laboratory. Details on using the various laboratories are found elsewhere in the Virtual ChemLab Help.

The Virtual ChemLab Environment

Given below is a brief tour of some of the important virtual laboratory interface elements that will be used heavily to get around in the laboratory and also a brief description of the types of experiments that can be performed in each of the laboratories.

Main Laboratory. Once the Virtual ChemLab simulation has started, the user is placed in the main laboratory, which contains various lab benches and a whiteboard. You can gain access to any of the laboratories by clicking on the

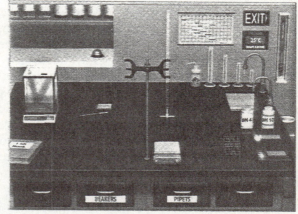

appropriate lab bench, which will then launch a holographic projector containing a close-up of the selected bench. You can also click on the periodic table located on the wall and the whiteboard for access to a list of preset experiments.

Holographic Projectors. As the user navigates through the virtual laboratory, holographic projectors are used to display close-ups of the various laboratories. The holographic window can be moved around the lab by clicking on and dragging the bar at the top of the holographic window.

Whiteboard and Worksheets. The whiteboard is used to display a list of worksheet assignments that are used with accompanying workbooks. While the laboratory is open-ended and is designed

7

to allow students to explore and perform any experiments they choose, in most cases students will be given worksheet assignments either online through a course management system, as PDF files, or on actual paper. In all of these cases, the worksheets provide step-by-step instructions to guide the student through an experiment, and the worksheets also require students to provide experimental data, answers to questions, and conclusions. In some cases when these worksheets are given the student will enter Virtual ChemLab bypassing the main room and will be placed at a specific laboratory. In other cases, however, students will be given a collection of worksheets and will be directed to perform a specific assignment. In this instance, the student will need to go to the whiteboard, select the worksheet page, and then locate and click on the appropriate assignment title. This will automatically bring the student to the appropriate laboratory for the assigned worksheet and set up the laboratory with the necessary equipment or configuration. Consequently, in many cases the whiteboard will be the main conduit for performing laboratory assignments using Virtual ChemLab.

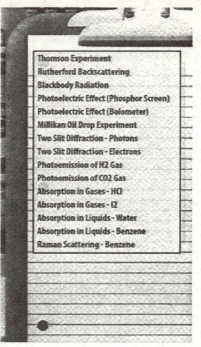

Preset Experiments or Clipboards. The Virtual ChemLab environment is designed to allow students access to the full range experiments available in each laboratory. While conceptually this is very powerful, in practice learning how to navigate all of the virtual laboratories and perform experiments can be intimidating and time consuming. The virtual laboratory mitigates this difficulty using preset experiments where each laboratory can be pre-configured for a certain experiment or limited to a smaller set of options. These preset experiments are located on clipboards in each of the laboratories and contain a list of preset titles. Mousing over a title pops up a short summary of the preset below the preset list. In every way, these presets are identical to the worksheet assignments listed on the whiteboard, however in the case of the clipboards these presets are not associated with specific worksheet assignments but are designed for more exploratory or general use.

When a preset has been selected in a particular laboratory, the laboratory will automatically be configured as defined by the preset and the lab book will have a blinking spotlight. Clicking on the lab book will open the lab book window and display any laboratory instructions or background information associated with the preset experiment. As data is recorded in the lab book or other links are shown, these instructions can always be displayed by clicking on the Preset menu button on the lab book. When a preset experiment is no longer needed, exiting the lab will generally clear the experiment or click on the Finish button located at the bottom of the preset instructions.

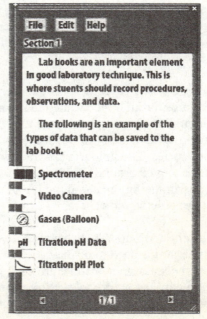

Lab Book. As in real laboratories, the lab book is used to record experimental observations and data. Clicking on the lab book in any of the laboratories will open the lab book window, which can be sized and moved within the laboratory window. All of the laboratories allow you to save experiment data to the lab book including gas data, titration data, calorimetry data, detector results from the quantum lab, and pictures from the inorganic lab. This

8

data is saved in the lab book as links, which can be clicked to open a Link window attached to the lab book that is used to show and in some cases analyze the saved data. Note that the lab book is not saved automatically and will be lost when you leave the virtual laboratory. However, the lab book can be saved locally and opened in a later laboratory session for further work and analysis. Details on using the lab book can be found in the Lab Book Help menu.

The following is a brief description of the capabilities of the five laboratories available in Virtual ChemLab.

Calorimetry. The calorimetry laboratory provides students with three different calorimeters that allow them to measure various thermodynamic processes including heats of combustion, heats of solution, heats of reaction, the heat capacity, and the heat of fusion of ice. The calorimeters provided in the simulations are a classic "coffee cup" calorimeter, a dewar flask (a better version of a coffee cup), and a bomb calorimeter. The calorimetric method used in each calorimeter is based on measuring the temperature change associated with the different thermodynamic processes. Students can choose from a wide selection of organic materials to measure the heats of combustion; salts to measure the heats of solution; acids, bases, oxidants, and reductants for heats of reaction; metals and alloys for heat capacity measurements; and ice for a melting process. Temperature versus time data can be graphed during the measurements and saved to the electronic lab book for later analysis. Systematic and random errors in the mass and volume measurements have been included in the simulation by introducing buoyancy errors in the mass weighings, volumetric errors in the glassware, and characteristic systematic and random errors in the thermometer measurements.

Gases. The gases laboratory allows students to explore and better understand the behavior of ideal gases, real gases, and van der Waals gases (a model real gas). The gases laboratory contains four experiments each of which includes the four variables used to describe a gas: pressure (P), temperature (T), volume (V), and the number of moles (n). The four experiments differ by allowing one of these variables to be the dependent variable while the others are independent. The four experiments include (1) V as a function of P, T, and n using a balloon to reflect the volume changes; (2) P as a function of V, T, and n using a motor driven piston; (3) T as a function of P, V, and n again using a motor driven piston; and (4) V as a function of P, T, and n but this time using a frictionless, massless piston to reflect volume changes and using weights to apply pressure. The gases that can be used in these experiments include an ideal gas; a van der Waals gas whose parameters can be changed to represent any real gas; real gases including N_2, CO_2, CH_4, H_2O, NH_3, and He; and eight ideal gases with different molecular weights that can be added to the experiments to form gas mixtures.

Quantum. The purpose of the quantum laboratory is to allow students to explore and better understand the foundational experiments that led to the development of quantum mechanics. Because of the very sophisticated nature of most of these experiments, the quantum laboratory is the most "virtual" of the Virtual ChemLab laboratory simulations. In general, the laboratory consists of an optics table where a source, sample, modifier, and detector combination can be placed to perform different experiments. These devices are located in the stockroom and can be taken out of the stockroom and placed in various locations on the optics table. The emphasis here is to teach students to probe a sample (e.g., a gas, metal foil, two-slit screen, etc.) with a source (e.g., a laser, electron gun, alpha-particle source, etc.) and detect the outcome with a specific detector (e.g., a phosphor screen, spectrometer, etc.). Heat, electric fields, or magnetic fields can also be applied to modify an aspect of the experiment. As in all the virtual laboratories, the focus is to allow students the ability to explore and discover, in a safe and level-appropriate setting, the concepts that are important in the development of atomic theory.

Titration. The virtual titration laboratory allows students to perform precise, quantitative titrations involving acid-base and electrochemical reactions. The available laboratory equipment consists of a 50 mL buret, 5, 10, and 25 mL pipets, graduated cylinders, beakers, a stir plate, a set of 8 acid-base indicators, a pH meter/voltmeter, a conductivity meter, and an analytical balance for weighing out solids. Acid-base titrations can be performed on any combination of mono-, di-, and tri-protic acids and mono-, di-, and tri-basic bases. The pH of these titrations can be monitored using a pH meter, an indicator, and a conductivity meter as a function of volume, and this data can be saved to an electronic lab book for later analysis. A smaller set of potentiometric titrations can also be performed. Systematic and random errors in the mass and volume measurements have been included in the simulation by introducing buoyancy errors in the mass weighings, volumetric errors in the glassware, and characteristic systematic and random errors in the pH/voltmeter and conductivity meter output. These errors can be ignored, which will produce results and errors typically found in high school or freshman-level laboratory work, or the buoyancy and volumetric errors can be measured and included in the calculations to produce results better than 0.1% in accuracy and reproducibility.

Inorganic. The features of the inorganic simulation include 26 cations that can be added to test tubes in any combination, 11 reagents that can be added to the test tubes in any sequence and any number of times, necessary laboratory manipulations, a lab book for recording results and observations, and a stockroom for creating test tubes with known mixtures, generating practice unknowns, or retrieving instructor assigned unknowns. The simulation uses over 2500 actual pictures to show the results of reactions and over 220 videos to show the different flame tests. Cations that are available in the simulation include Ag^+, Al^{3+}, Ba^{2+}, Bi^{3+}, Ca^{2+}, Cd^{2+}, Co^{2+}, Cr^{3+}, Cu^{2+}, Fe^{2+}, Fe^{3+}, Hg^{2+}, Hg_2^{2+}, K^+, Mg^{2+}, Mn^{2+}, Na^+, NH_4^+, Ni^{2+}, Pb^{2+}, Sb^{3+}, Sn^{4+}, Sr^{2+}, Ti^{4+}, V^{4+}, and Zn^{2+}. Available reagents include HNO_3, H_2O_2, NH_3, Na_2S, $NaCl$, Na_2SO_4, $NaOH$, Na_2CO_3 and pH 4, pH 7, and pH 10 buffers. With 26 cations that can be combined in any order or combination and 11 reagents that can be added in any order, there are in excess of 10^{16} possible outcomes in the simulation.

Important Installation Notes and Issues

1. The graphics used in the simulations require the monitor to be set to 24-bit true color (millions of colors). Lower color resolutions can be used, but the graphics will not be as sharp.

2. Right clicking anywhere in the lab will bring up several menu options, one of which shows the current software and database version numbers. On some Macintosh computers, after exiting this menu and returning to the lab tooltips, rollovers, and highlighting will no longer work. This issue is fixed by exiting and restarting the program.

3. On some monitors at expanded resolutions some of the graphics will appear to have pixel offsets at random locations within the laboratory window. There is no known cause for this, but it can be fixed by moving the laboratory window off from center within the main laboratory window.

4. On the clipboard in the Titration stockroom, there are several different unknowns that can be selected by students. Each unknown can be assigned with unknown numbers ranging from 1 to 15. In rare cases, an unknown number 16 can be assigned, which is not a valid unknown number resulting in unpredictable results. The unknown numbers are randomly seeded using a combination of variables, one of which is the username used to log into the computer. Logging in with a different username will fix the problem.

12

Workbook Assignments

The following laboratory assignments cover most of the topics taught in the general chemistry curriculum. The purpose of these assignments is to allow you to put into practice the concepts and problem solving skills you have been learning in the classroom. Some of these assignments merely allow you to measure and collect the data that would normally be provided for you in a homework problem. Other assignments allow you to perform sophisticated and foundational experiments to which you would not normally have access, while other assignments allow you to perform experiments typically available in the undergraduate laboratory but much more quickly and cleanly. In all, you will find that these laboratory assignments using the virtual laboratory will provide a bridge in understanding between the abstract concepts of the classroom and their application in the actual laboratory.

For each laboratory assignment provided in this workbook, there is a corresponding assignment listed on the whiteboard located in the front of the laboratory. Clicking on an assignment on the whiteboard will project the appropriate laboratory bench in the general chemistry laboratory and setup the lab bench for the selected experiment. Most assignments should take no longer than 15 to 20 minutes, especially once you have learned how to use *Virtual ChemLab*. Remember, the purpose of these experiments is to give you practice in thinking, problem solving, and applying concepts. Feel free to setup experiments and explore in the virtual laboratory. There are, essentially, an unlimited number of things you can discover.

14

1-1: Thomson Cathode Ray Tube Experiment

As scientists began to examine atoms, their first discovery was that they could extract negatively charged particles from atoms. They called these particles electrons. In order to understand the nature of these particles, scientists wanted to know how much charge they carried and how much they weighed. John Joseph (J.J.) Thomson was a physics professor at the famous Cavendish Laboratory at Cambridge University. In 1897, Thomson showed that if you could measure how far a beam of electrons was bent in an electric field and in a magnetic field, you could determine the charge-to-mass ratio (q/m_e) for the particles (electrons). Knowing the charge-to-mass ratio (q/m_e) and either the charge on the electron or the mass of the electron would allow you to calculate the other. Thomson could not obtain either in his cathode ray tube experiments and had to be satisfied with just the charge-to-mass ratio.

1. Start *Virtual ChemLab*, select *Atomic Theory*, and then select *Thomson Cathode Ray Tube Experiment* from the list of assignments. The lab will open in the Quantum laboratory.

2. *What source is used in this experiment?* Drag your cursor over to the source to identify it._____

   _____

   *What type of charge do electrons have?*_____

   What detector is used in this experiment? _____

3. Turn on the *Phosphor Screen* by clicking on the red/green light switch.

   *What do you observe?*_____

   The phosphor screen detects charged particles (such as electrons) and it glows momentarily at the positions where the particles impact the screen.

4. It may be helpful to drag the lab window down and left and the phosphor screen window up and right in order to minimize overlap. Push the **Grid** button on the phosphor screen, and set the *Magnetic Field* to 30 μT. (Click buttons above and below the digits in the meter to raise and lower the value. Clicking between digits moves the decimal point.)

   What happens to the spot from the electron gun on the phosphor screen? _____

   _____

5. Set the *Magnetic Field* back to zero and set the *Electric Field* to 10 V.

   *What happens to the spot on the phosphor screen?*_____

   Where should the signal on the phosphor screen be if the electric and magnetic forces are balanced?

   _____

6. Increase the voltage of the *Electric Field* so the spot is 5 cm left of center.

   *What voltage is required?*_____

7. Increase the magnetic field strength until the spot reaches the center of the screen.

   What magnetic field creates a magnetic force that balances the electric force? _____

   ### Summarize your data.

deflected distance (*d*)	electric field (*V*)	magnetic field (*B*)

8. In a simplified and reduced form, the charge-to-mass ratio (q/m_e) can be calculated as follows:

   $$q/m_e = \left(5.0826 \times 10^{12}\right) \cdot V \cdot d / B^2$$

   where V = the electric field in volts, d = the deflected distance from center in cm after applying just the voltage, and B = magnetic field in μT.

   What is your calculated value for the charge-to-mass ratio for an electron (q/m_e)? _____

   _____

   The modern accepted value is 1.76×10^{11}.

   Calculate your percent error as follows:

   $$\% \ Error = \frac{|your \ value - accepted \ value|}{accepted \ value} \times 100$$

   % Error = _____

9. You may want to repeat the experiment several times using different size deflections.

1-2: Millikan Oil Drop Experiment

In the Thomson Cathode Ray Tube Experiment, it was discovered that you can use the deflection of an electron beam in an electric and magnetic field to measure the charge-to-mass ratio (q/m_e) of an electron. If you then want to know either the charge or the mass of an electron, you need to have a way of measuring one or the other independently. In 1909, Robert Millikan and his graduate student Harvey Fletcher showed that they could make very small oil drops and deposit electrons on these drops (1 to 10 electrons per drop). They would then measure the total charge on the oil drops by deflecting the drops with an electric field. You will get a chance to repeat their experiments and, using the results from the Thomson assignment, be able to experimentally calculate the mass of an electron.

1. Start *Virtual ChemLab,* select *Atomic Theory,* and then select *Millikan Oil Drop Experiment* from the list of assignments. The lab will open in the Quantum laboratory.

2. *What is the purpose of the electron gun in this experiment?* _____

   _____

   How does this source affect the oil droplets in the oil mist chamber? _____

   _____

3. The detector in this experiment is a video camera with a microscopic eyepiece attached to view the oil droplets. Click the *On/Off* switch (red/green light) to turn the video camera on.

   What do you observe on the video camera screen? _____

   _____

   Do all the oil drops fall at the same speed? _____

   What force causes the drops to fall? _____

   The oil drops fall at their terminal velocity, which is the maximum velocity possible due to frictional forces such as air resistance. The terminal velocity is a function of the radius of the drop. By measuring the terminal velocity (v_t) of a droplet, the radius (r) can be calculated. Then the mass (m) of the drop can be calculated from its radius and the density of the oil. Knowing the mass of the oil droplet will allow you to calculate the charge (q) on the droplet.

   IMPORTANT: Read instructions 4 and 5 before beginning the procedure for 5.

4. *Measure the terminal velocity of a drop.* Identify a small drop near the top of the window that is falling near the center scale and click the *Slow Motion* button on the video camera. Wait until the drop is at a tick mark and start the timer. Let the drop fall for at least two or more tick marks and stop the timer. Do not let the drop fall off the end of the viewing scope. Each tick mark is 0.125 mm. Record the distance and the time in the data table on the following page.

5. *Measure the voltage required to stop the fall of the drop.* Having measured the terminal velocity, you now need to stop the fall of the drop by applying an electric field between the two voltage plates. This

is done by clicking on the buttons on the top or bottom of the *Electric Field* until the voltage is adjusted such that the drop stops falling. This should be done while in slow motion. When the drop appears to stop, turn the slow motion off and do some final adjustments until the drop has not moved for at least one minute. Record the voltage, V, indicated on the voltage controller.

Complete the experiment for three drops and record your measurements in the data table.

Data Table

Drop	Voltage (V, in volts)	Time (t, in seconds)	Distance (d, in meters)
1			
2			
3			

The *Millikan Oil Drop Experiment* is a classic due to the simplicity of the experimental apparatus and the completeness of the data analysis. The following calculations have reduced very complex equations into simpler ones with several parameters combined into a single constant. Millikan and Fletcher accounted for the force of gravity, the force of the electric field, the density of the oil, the viscosity of the air, the viscosity of the oil, and the air pressure.

6. *Calculate the terminal velocity and record the value.* Calculate the terminal velocity, v_t, in units of m·s^{-1} using this equation:

    $v_t = \dfrac{d}{t}$, where d is the distance the drop fell in meters and t is the elapsed time in seconds. Do not forget that the scale on the viewing scope is in mm (1000 mm = 1 m). _____

    _____

    Each of the equations in instructions 7-10 is shown with and without units. You will find it easier to use the equation without units for your calculations.

7. *Calculate the radius (r) of the drop and record the value.* With the terminal velocity, you can calculate the radius, in m, of the drop using this equation:

    $$r = \left(9.0407 \times 10^{-5}\,\text{m}^{1/2} \cdot \text{s}^{1/2}\right) \cdot \sqrt{v_t} \quad = (9.0407 \times 10^{-5} \sqrt{v_t}\ , \text{without units})$$

    _____

8. *Calculate the mass of the drop and record the value.* You can use the answer from #7 for the radius (r) to calculate the mass of the drop given the density of the oil. The final equation to calculate the mass, in kg, is

    $$m = V_{oil} \cdot \rho_{oil} = 4\pi/3 \cdot r^3 \cdot 821\text{kg} \cdot \text{m}^{-3}$$
    $$= \left(3439.0\text{kg} \cdot \text{m}^{-3}\right) \cdot r^3 \qquad = (3439.0 \cong r^3\ , \text{without units})$$

    _____

9. Since you applied a voltage across the *Electric Field* to stop the fall of the oil drop, the forces being exerted on the drop must be balanced; that is, the force due to gravity must be the same as the force due to the electric field acting on the electrons stuck to the drop: $qE = mg$.

   Calculate the total charge (Q_{tot}) on the oil drop due to the electrons using the equation:

   $$Q_{tot} = Q(n) \cdot e = \left(9.810 \times 10^{-2} \, C \cdot kg^{-1} \cdot J^{-1}\right) \cdot m/V \ = (9.81 \times 10^{-2} \, m/V, \text{ without units})$$

   where $Q(n)$ is the number of electrons on the drop, e is the fundamental electric charge of an electron, m is the mass calculated in #8, and V is the voltage.

   This answer will provide the total charge on the drop (Q_{tot}). The fundamental electric charge of an electron (e) is 1.6×10^{-19} C (coulombs). Divide your total charge (Q_{tot}) by e and round your answer to the nearest whole number. This is the number of electrons ($Q(n)$) that adhered to your drop. Now divide your total charge (Q_{tot}) by $Q(n)$ and you will obtain your experimental value for the charge on one electron.

10. Complete the experiment and calculations for at least three drops and summarize your results in the results table.

Results Table

Drop #	Terminal Velocity (v_t, in m/s)	Radius (r, in meters)	Mass (m, in kg)	Total charge on drop (Q_{tot}, in Coulombs)	Charge on one electron (C)
1					
2					
3					

11. Average your results for the charge on one electron. Calculate the percent error by:

$$\% \ Error = \frac{\left| your \ answer - 1.6 \times 10^{-19} \right|}{1.6 \times 10^{-19}} \times 100\%$$

What is your average charge for an electron? _____

_____

What is your percent error? _____

12. You will recall that in the Thomson experiment you were able to calculate the charge-to-mass ratio (q/m_e) as 1.7×10^{11}. Using this value for q/m_e and your average charge on an electron, calculate the mass of an electron in kg.

What is your calculated value for the mass of an electron in kg? _____

_____

1-3: Rutherford's Backscattering Experiment

A key experiment in understanding the nature of atomic structure was completed by Ernest Rutherford in 1911. He set up an experiment that directed a beam of alpha particles (helium nuclei) through a gold foil and then onto a detector screen. According to the "plum-pudding" atomic model, electrons float around inside a cloud of positive charge. Based on this model, Rutherford expected that almost all of the alpha particles should pass through the gold foil and not be deflected. A few of the alpha particles would experience a slight deflection due to the attraction to the negative electrons (alpha particles have a charge of +2). Imagine his surprise when a few alpha particles deflected at all angles, even nearly straight backwards.

According to the "plum-pudding" model there was nothing in the atom massive enough to deflect the alpha particles. Rutherford's reaction was that this was ". .almost as incredible as if you fired a 15-inch shell at a piece of tissue paper and it came back and hit you." He suggested that the experimental data could only be explained if the majority of the mass of an atom was concentrated in a small, positively charged central nucleus. This experiment provided the evidence needed to prove the nuclear model of the atom. In this experiment, you will make observations similar to those of Professor Rutherford.

1. Start *Virtual ChemLab,* select *Atomic Theory,* and then select *Rutherford's Backscattering Experiment* from the list of assignments. The lab will open in the Quantum laboratory.

2. The experiment will be set up on the lab table. The gray box on the left side of the table contains a sample of ^{241}Am.

    What particles are emitted from this source? _____

    What are alpha particles? _____

    _____

3. Mouse over the metal foil stand in the middle of the table.

    What metal foil is used? If you want to see the metal foil, click and hold on the metal stand. _____

4. Point the cursor to the detector (on the right).

    What detector is used in this experiment? _____

5. Turn on the detector by clicking on the red/green light switch.

    What does the signal in the middle of the screen represent? _____

    _____

    The phosphor screen detects charged particles (such as alpha particles) and it glows momentarily at the positions where the particles impact the screen.

    What other signals do you see on the phosphor detection screen? _____

    _____

What do these signals represent? _____

Click the **Persist** button (the dotted arrow) on the phosphor detector screen.

According to the plum pudding model, what is causing the deflection of the alpha particles?

_____

_____

Make a general observation about the number of alpha particles that hit the phosphor detection screen in a minute's worth of time.

_____

6. Now, you will make observations at different angles of deflection. Click on the main laboratory window to bring it to the front. Grab the phosphor detection screen by its base and move it to the spotlight in the top right corner. The **Persist** button should still be on.

   Observe the number of hits in this position as compared with the first detector position. _____

   _____

7. Move the detector to the top center spotlight position at a 90E angle to the foil stand.

   Observe the number of hits in this spotlight position as compared with the first detector position. ___

   _____

8. Move the detector to the top left spotlight position and observe the number of hits on the phosphor screen in one minute.

   Observe the number of hits in this spotlight position as compared with the first detector position. ___

   _____

   What causes the alpha particles to deflect backwards? _____

   How do these results disprove the plum pudding model? Keep in mind that there are 1,000,000 alpha particles passing through the gold foil per second.

   _____

   _____

   Are the gold atoms composed mostly of matter or empty space? _____

   How does the Gold Foil Experiment show that almost all of the mass of an atom is concentrated in a small positively charged central atom?

   _____

_____

_____

Students often ask, "Why did Rutherford use gold foil?" The most common response is that gold is soft and malleable and can be made into very thin sheets of foil. There is another reason, which you can discover for yourself.

9. Turn off the phosphor detection screen. Click and drag the base of the metal foil holder to the stockroom counter. Click on the *Stockroom* to enter. Click on the metal sample box on the top shelf. Click on *Mg* to select magnesium. Click on the *Return to Lab* arrow.

10. Move the metal foil sample holder from the stockroom window back to the center of the table. Move the phosphor screen back to its original location on the right side of the table and turn it on. Click *Persist*. Observe the number of hits with magnesium compared with the number of hits with a gold sample.

    Why would Rutherford choose gold foil instead of magnesium foil? Explain. _____

_____

_____

_____

1-4: Investigating the Properties of Alpha and Beta Particles

As scientists began investigating the properties of atoms, their first discovery was that they could extract negatively charged particles. They called these particles electrons, but they are also known as beta particles in the context of nuclear decay. Robert Millikan used beta particles in his famous Oil Drop Experiment. Another particle ejected during nuclear decay is the alpha particle. An alpha particle is a helium nucleus, or a helium atom without its two electrons. Consequently, an alpha particle is positively charged. Ernest Rutherford used alpha particles in his Gold Foil Experiment.

1. Start *Virtual ChemLab,* select *Atomic Theory,* and then select *Alpha and Beta Particles* from the list of assignments. The lab will open in the Quantum laboratory.

2. *What source is used in this experiment?* Drag your cursor over the source to identify it. _____

   What type of charge do electrons have? _____

   What detector is used in this experiment? _____

3. Turn on the *Phosphor Screen.* (Click on the green/red button.)

   What do you observe? _____

   The phosphor screen detects charged particles (such as electrons) and it glows momentarily at the positions where the particles impact the screen.

4. Drag the lab window down and left and the phosphor screen window up and right in order to minimize the overlap. Push the **Grid** button on the phosphor screen, and set the *Magnetic Field* to 30 μT. (Click the button above the tens place three times. If you mistakenly click between digits, it will move the decimal point. Click it to place it where it was originally and then click above the tens place.)

   What happens to the spot from the electron gun on the phosphor screen? _____

5. Click once above the tens place on the *Electric Field* meter. Observe the spot. Click a second time above the tens place on the *Electric Field.*

   What happens to the spot from the electron gun on the phosphor screen? _____

6. Zero out the *Magnetic Field* and *Electric Field* meters by clicking on the appropriate digit buttons until the spot on the phosphor screen is centered again.

7. Click and drag the electron gun to move it to the *Stockroom* counter. Enter the *Stockroom* by clicking inside the *Stockroom.* Double-click the electron gun to move it back to the shelf. Double-click on the alpha source to select it and move it to the *Stockroom* counter. Click on the green *Return to Lab* arrow to return to the lab. Drag the alpha source from the *Stockroom* counter and place it on the table where the electron gun was originally placed (the middle spot light). Click on the front of the alpha source to open the shutter.

   What appears on the phosphor screen? _____

8. Change the unit for the *Magnetic Field* from µT to mT by clicking once above the unit. Click above the hundreds place three times to set the *Magnetic Field* to 300 mT (millitesla). This magnetic field is one million times stronger than what we used for the electron gun.

   Which direction is the spot deflected when the magnetic field is increased this time? _____

   How does this compare with the direction of movement when the magnetic field was turned on for the electrons?

   _____

   _____

9. Change the unit for the *Electric Field* from V to kV by clicking once above the unit. Observe the spot as you increase the *Electric Field* strength from 0 kV to 5 kV. The movement is slight so pay careful attention.

   Which direction is the spot moved when you increase the Electric Field? _____

   How does this compare with the direction of movement for the electron beam in the Electric Field?

   _____

   Why does it take significantly stronger magnetic and electric field strengths to move the beam of alpha particles compared with the beam of electrons (beta particles)?

   _____

10. Return the values on the two meters to zero. Click and drag the alpha source and the phosphor screen to return them to the *Stockroom* counter. Enter the *Stockroom*. Double-click on the alpha source and the phosphor screen to place them on the shelf. Select the laser and the video camera by double-clicking on them, and then click on the green *Return to Lab* arrow to return to the lab. Place the laser in the center spotlight on the left and turn the laser on (click on the red/green light). Place the video camera on the center spotlight on the right and click the video camera to turn it on. Set the laser intensity to 1 nW and the wavelength to 20 nm. This wavelength is in the x-ray region of the electromagnetic spectrum. The purple dot is a representation of the x-rays hitting the video camera. Change the *Electric Field* and *Magnetic Field* to determine the effect on the x-rays.

    Did the electric or magnetic fields affect the x-rays? Why or why not? _____

    _____

    Summarize what you have learned about electrons (beta particles), alpha particles, and x-rays.

    _____

    _____

    _____

1-5: Blackbody Radiation

In the early 1900s several experimental results appeared to be in conflict with classical physics. One of these experiments was the study of blackbody radiation. A blackbody is a solid (such as a piece of iron) that does not emit light at low temperatures, but, when heated, the "blackbody" begins to emit first red and then orange light and at higher temperatures eventually becomes white hot. The intensity of the emitted light is also a function of temperature. In this problem you will make observations similar to those of Max Planck (1858-1947) who, through his study of blackbody radiation, found an explanation that revolutionized how scientists think about radiated energy.

1. Start *Virtual ChemLab*, select *Atomic Theory*, and then select *Blackbody Radiation* from the list of assignments. The lab will open in the Quantum laboratory. A metal sample holder with tungsten metal will be on the lab bench with an electric heater set at a temperature of 3000 K. A spectrometer is on the right and is switched on (the spectrometer window is open). Locate the switch that changes the display from wavelength to frequency and the switch that displays either the full electromagnetic spectrum or just the visible spectrum. These will be used later.

2. The spectrometer detects the intensity of the emitted light as a function of the wavelength (or frequency). In the grid below draw the spectrum detected by the spectrometer with wavelength (in nm) on the x-axis and intensity on the y-axis. If you drag your cursor over a peak, it will identify the wavelength (in nm) in the x-coordinate field in the bottom right corner of the detector window. Record the wavelength of the peak in the data table on the following page. (Round to whole numbers.)

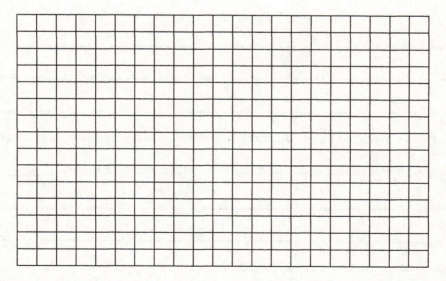

3. Change the temperature on the heater to 3100 K by clicking on the button above the hundreds place on the heater LCD controller. Record the shape of the curve on the same graph (label each line with a temperature) and the wavelength of the peak intensity in the data table. Continue with temperatures of 3200 K, 3300 K, 3400 K, 3500 K, and 3600 K. If you raise the temperature to 3700 K you will have to start over by clicking on the *Reset Lab* button just under the Danger sign, entering the *Stockroom*, clicking on the clipboard and selecting Preset Experiment #3 Blackbody Radiation.

Data Table

Temperature (K)	Wavelength (nm)
3000	
3100	
3200	
3300	
3400	
3500	
3600	

4. *What observations can you make about the shape of the curve as temperature changes?*

_____

_____

5. The visible portion of the electromagnetic spectrum occurs between 400 nm and 700 nm. Mark the visible spectrum on your graph.

   Does the peak intensity ever occur in the visible region? Does this mean that there is no visible light

   radiated over this temperature range? Explain. _____

   _____

   _____

   _____

   _____

6. Classical physics predicted that as the wavelength decreases the intensity should increase.

   Does your graph of wavelength and intensity confirm this result? Explain. _____

   _____

   _____

7. In the spectrometer window, change the display from wavelength to frequency.

   What observations can you make about the magnitude of the intensity as the temperature is lowered

   from 3600 K to 3000 K by increments of 100 K? _____

   _____

8. Change the temperature to 3700 K while observing the tungsten foil.

   What occurs at 3700 K? _____

9. This disparity between classical theory and experimental results was known as the "ultraviolet catastrophe." Classical physics predicted that the curve should continue to infinity as the wavelength is decreased. The conflicting experimental data required an innovative explanation. This explanation was provided by Max Planck who stated that the energy given off by the vibrating heated atoms was *quantized* and only vibrations of specific energy could occur. The quantized energy must be a multiple of $h\nu$ where h is known as Planck's constant (6.626×10^{-34} J·s) and ν is the frequency of the light in 1/s or s^{-1}. Calculate the energy of the peak intensity at 3400 K.

Use the wavelength from the data table for 3400 K to determine first the frequency ($\nu = c/\lambda$, where c = speed of light = 2.998×10^8 m·s^{-1} and λ is the wavelength in meters) and then find the energy using $E = h\nu$.

1-6: Photoelectric Effect

Although Albert Einstein is most famous for $E = mc^2$ and for his work describing relativity in mechanics, his Nobel Prize was for understanding a very simple experiment. It was long understood that if you directed light of a certain wavelength at a piece of metal, it would emit electrons. In classical theory, the energy of the light was thought to be based on its intensity and not its frequency. However, the results of the photoelectric effect contradicted classical theory. Inconsistencies led Einstein to suggest that we need to think of light as being composed of particles (photons) and not just as waves. In this experiment, you will reproduce a photoelectric experiment and show that the energy (E) of a photon of light is related to its frequency and not its intensity.

1. Start *Virtual ChemLab,* select *Atomic Theory,* and then select *Photoelectric Effect* from the list of assignments. The lab will open in the Quantum laboratory.

2. *What source is used in this experiment and what does it do?* _____

    _____

    At what intensity is the laser set? _____

    At what wavelength is the laser set? _____

    Record the wavelength (in nm) in the data table on the following page. Calculate the frequency (in Hz) and the energy (in J) using $c = \lambda v$ and $E = hv$ where $c = 2.998 \times 10^8$ m·s^{-1} and $h = 6.626 \times 10^{-34}$ J·s. Also record the color of the light by clicking on the *Spectrum Chart* (just behind the laser); the marker indicates what color is represented by the wavelength selected.

    Which metal foil is used in this experiment? _____

    What detector is used in this experiment and what does it measure? _____

    _____

    Turn on the detector by clicking on the red/green light switch.

    What does the signal on the phosphor screen indicate about the laser light shining on the sodium

    foil? _____

3. *Decrease the Intensity to 1 photon/second, how does the signal change?* _____

    _____

    Increase the Intensity to 1kW, how does the signal change? _____

    _____

    Change the *Intensity* back to 1 nW and increase the *Wavelength* to 600 nm.

    What do you observe? Record the wavelength in the data table. _____

Determine the maximum wavelength at which emission of electrons occurs in the metal. _____

What is the difference between intensity and wavelength? _____

_____

Which matters in the formation of photoelectrons: intensity or wavelength? _____

_____

Data Table

wavelength (nm)	frequency (1/s)	energy (J)	light color

4. Click inside the *Stockroom* to enter the stockroom. Click on the clipboard and select the preset experiment called *Photoelectric Effect (Bolometer)*. Click on the green *Return to Lab* arrow to return to the laboratory. The intensity of the laser will be set at 1 nW and the wavelength at 400 nm. The detector used in this experiment is a bolometer and will be automatically turned on. This instrument measures the kinetic energy of electrons. You should see a green peak on the bolometer detection screen. The intensity or height of the signal corresponds to the number of electrons being emitted from the metal, and the *x*-axis is the kinetic energy of the electrons. Zoom in on the peak by clicking and dragging from the left of the peak to the right.-

5. *Increase and decrease the Intensity, what do you observe?* _____

_____

Increase and decrease the Wavelength, what do you observe? _____

_____

_____

What is the maximum wavelength that ejects electrons from the sodium metal? _____

Based on this experiment, explain why violet light causes photoemission of electrons but orange light

does not. _____

_____

1-7: The Rydberg Equation

When a sample of gas is excited by applying a large alternating electric field, the gas emits light at certain discrete wavelengths. In the late 1800s two scientists, Johann Balmer and Johannes Rydberg, developed an emperical equation that correlated the wavelength of the emitted light for certain gases such as H_2. Later, Niels Bohr's concept of quantized "jumps" by electrons between orbits was shown to be consistent with the Rydberg equation. In this assignment, you will measure the wavelengths of the lines in the hydrogen emission spectra and then graphically determine the value of the Rydberg constant, R_H.

1. Start *Virtual ChemLab,* select *Atomic Theory,* and then select *The Rydberg Equation* from the list of assignments. The lab will open in the Quantum laboratory. The *Spectrometer* will be on the right of the lab table. The hydrogen emission spectra will be in the detector window in the upper right corner as a graph of intensity vs. wavelength (λ).

2. *How many distinct lines do you see and what are their colors?* _____

    _____

3. Click on the ***Visible/Full*** switch to magnify only the visible spectrum. You will see four peaks in the spectrum. If you drag your cursor over a peak, it will identify the wavelength (in nm) in the *x*-coordinate field in the bottom right corner of the detector window. Record the wavelengths of the four peaks in the visible hydrogen spectrum in the data table. (Round to whole numbers.)

4. The Rydberg equation has the form $\dfrac{1}{\lambda} = R_H \left(\dfrac{1}{n_f^2} - \dfrac{1}{n_i^2} \right)$ where λ is the wavelength in meters, R_H is the

    Rydberg constant, n_f is the final principal quantum (for the Balmer series, which is in the visible spectrum, $n_f = 2$), and n_i is the initial principal quantum number ($n = 3, 4, 5, 6, ..$). Calculate from your experimental data the wavelength in meters and $1/\lambda$ in m^{-1}. Record your answers in the data table.

Data Table

	λ (nm)	λ (m)	$1/\lambda$ (m^{-1})
Line #1 (left)			
Line #2			
Line #3			
Line #4 (right)			

5. The formula for the determination of energy is $E = h\nu = hc/\lambda$ where h is Planck's constant and c is the speed of light. *What is the relationship between wavelength and energy?*

    _____

6. *Of the four measured hydrogen spectrum lines recorded on the previous page, which line corresponds to the transition n = 3 to n = 2, and from n = 4 to n = 2, and so on from n = 6 to n = 2?*

_____

_____

_____

7. Calculate the value of $\left(\dfrac{1}{n_f^2} - \dfrac{1}{n_i^2}\right)$ for the transitions *n* = 6 to *n* = 2, *n* = 5 to *n* = 2, *n* = 4 to *n* = 2 and *n*

    = 3 to *n* = 2. Match the values for these transitions and record them with the appropriate reciprocal wavelength in the results table.

Results Table

Transition n_i to n_f	$\left(\dfrac{1}{n_f^2} - \dfrac{1}{n_i^2}\right)$	$1/\lambda$ (m^{-1})

8. The Rydberg equation, $\dfrac{1}{\lambda} = R_H\left(\dfrac{1}{n_f^2} - \dfrac{1}{n_i^2}\right)$, is in the form of $y = mx + b$ where $1/\lambda$ corresponds to y,

    $\left(\dfrac{1}{n_f^2} - \dfrac{1}{n_i^2}\right)$ corresponds to x, and $b = 0$. If you plot $1/\lambda$ on the y-axis and $\left(\dfrac{1}{n_f^2} - \dfrac{1}{n_i^2}\right)$ on the x-axis, the

    resulting slope will be the Rydberg constant, R_H.

    Using a spreadsheet program or a piece of graph paper, plot your experimental data and determine

    the value of the Rydberg constant. _____

9. The accepted value for R_H is 1.0974×10^7 m^{-1}.

    Determine the % error using the formula:

    $$\% \, Error = \frac{|your\ answer - accepted\ answer|}{accepted\ answer} \times 100$$

    % Error =

1-8: Atomic Emission Spectra

When a sample of gas is excited by applying a large alternating electric field, the gas emits light at certain discrete wavelengths. The intensity and wavelength of the light that is emitted is called the atomic emission spectrum and is unique for each gas. In this assignment, you will measure the emission spectra for several gases and then make some observations about the differences in their spectra.

1. Start *Virtual ChemLab,* select *Atomic Theory,* and then select *Atomic Emission Spectra* from the list of assignments. The lab will open in the Quantum laboratory. The *Spectrometer* will be on the right of the lab table. The hydrogen emission spectra will be in the detector window in the upper right corner as a graph of Intensity vs. wavelength (λ).

2. *How many distinct lines do you see and what are their colors?* _____

   _____

3. Click on the **Visible/Full** switch to magnify only the visible spectrum. You will see four peaks in the spectrum. If you drag your cursor over a peak, it will identify the wavelength (in nm) in the *x*-coordinate field in the bottom right corner of the detector window. Record the wavelengths of the four peaks in the visible hydrogen spectrum in the data table. (Round to whole numbers.)

4. The wavelength of each line can also be described in terms of its frequency. Use the wavelength of each line to calculate its frequency given that $v = c/\lambda$ where $c = 2.998 \times 10^{17}$ nm·s^{-1} (2.998×10^8 m·s^{-1}). The energy (E) of a single quantum of light emitted by an atom is related to its frequency (v) by the equation $E = hv$ where $h = 6.626 \times 10^{-34}$ J·s. Calculate the frequency of each line and the corresponding energy and record your results in the data Table.

Data Table

	λ (nm)	v (1/s)	Energy (J)
Line #1 (left)			
Line #2			
Line #3			
Line #4 (right)			

5. Now, investigate the emission spectra for a different element, helium. Helium is the next element after hydrogen on the periodic table and has two electrons.

   Do you think the emission spectra for an atom with two electrons instead of one will be much

   different than hydrogen? _____

6. To exchange gas samples, click and drag the *Electric Field* and place it on the stockroom counter, and double-click or click and drag the *Gas (H_2)* sample tube and place it on the stockroom counter as well. You may have to first click on the main laboratory window in order to move the items.

7. Enter the stockroom by clicking in the *Stockroom.* Click on the *Gases* samples on the top shelf. Click on the cylinder labeled *He* to replace the H_2 in the sample tube with helium gas. If you point to the gas sample tube with the cursor it should read *He*.

8. Return to the laboratory and drag the gas sample tube off the stockroom counter and place it in the middle of the table as indicated by the spotlight. Drag the *Electric Field* and place it on the gas sample tube. Carefully click the button just above the left zero on the *Electric Field* controller and change the voltage to 300 V. Turn on the *Spectrometer* by clicking on the red/green button and click the ***Visible/Full*** switch to view only the visible spectrum.

9. *Is this spectrum different than hydrogen? How many lines are present and what are their*

    colors? _____

10. *Determine the wavelength (in nm), the frequency (in 1/s) and the energy (in J) for the peak on the far right.*

	λ (nm)	v (1/s)	Energy (J)
Line (far right)			

1-9: Heisenberg Uncertainty Principle

It has long been known that if you shine light through narrow slits that are spaced at small intervals, the light will form a diffraction pattern. A diffraction pattern is a series of light and dark patterns caused by wave interference. The wave interference can be either constructive (light) or destructive (dark). In this experiment, you will shine a laser through a device with two slits where the spacing can be adjusted and then you will investigate the patterns that will be made at a distance from the slits.

1. Start *Virtual ChemLab*, select *Atomic Theory,* and then select *Heisenberg Uncertainty Principle* from the list of assignments. The lab will open in the Quantum laboratory.

2. *What source is used in this experiment and for what reason?* _____

    _____

    At what wavelength is the laser set? _____

    What is the spacing of the two slits on the two-slit device? _____

    Sketch a small picture of the pattern displayed on the video screen.

3. Change the *Intensity* of the laser from 1 nW to 1W.

    Does the intensity of the light affect the diffraction pattern? _____

    Change the *Slit Spacing* to 1μm. Observe the pattern displayed on the video screen as you change the slit spacing from 1 μm to 7 μm in increments of 1μm.

    What can you state about the relationship between slit spacing and diffraction pattern?

    _____

4. Increase the *Wavelength* of the laser to 700 nm.

    What effect does an increase in the wavelength have on the diffraction pattern? _____

    _____

5. Decrease the *Intensity* on the laser to 1000 photons/second. Click on the **Persist** button on the video camera to look at individual photons coming through the slits. Observe for one minute.

   What observation can you make about this pattern compared with the pattern from the continuous

   beam of photons? _____

   _____

   Decrease the *Intensity* to 100 photons/second. Observe for another minute after clicking **Persist**. At these lower intensities (1000 and 100 photons/second), there is never a time when two photons pass through the slits at the same time.

   How can a single photon diffract?

   _____

   _____

   _____

   _____

   Based on this experiment, what conclusions can you make about the nature of light?

   _____

   _____

6. Click inside the *Stockroom* to enter the stockroom. Click on the clipboard and select the preset experiment *Two Slit Diffraction – Electrons*. Click on the green *Return to Lab* arrow to return to the laboratory.

   What source is used in this experiment? _____

   Sketch a small picture of the diffraction pattern shown on the Phosphor Screen.

   How does this diffraction pattern compare with the diffraction pattern for light? _____

   _____

Louis de Broglie was the first person to suggest that particles could be considered to have wave properties.

7. Decrease the *Intensity* to 10 electrons/second. The pattern now builds one electron at a time. Click on the **Persist** button and observe for one minute.

    Has the diffraction pattern changed? Why or why not? _____

    _____

    How can a single electron diffract? _____

    _____

    _____

    _____

When you look at the complete diffraction pattern of a stream of particles, you are seeing all the places you expect particles to scatter. If you start the source over multiple times, you will see that the first particle is never detected in the same place twice. This is an application of the Heisenberg Uncertainty Principle, which is directly connected with measurement. It takes into account the minimum uncertainty of the position (Δx) and the uncertainty of the momentum (Δp) using the equation $(\Delta x)(\Delta p) \geq h/4\pi$. Because you know the energy at which the particle is traveling, you can precisely know the momentum, but you cannot know the position. Consequently, you cannot predict where each particle will hit.

1-10: Emission Spectra for Sodium and Mercury

In the 1800s, scientists found that when a sample of gas was excited by an alternating electric field, light with only certain discrete wavelengths was emitted. This property allowed for the development of spectroscopic techniques that can be used in the identification and analysis of elements and compounds. Even though scientists found spectroscopy very useful, they could not explain why the spectrum was not continuous. The explanation of this was left to Niels Bohr, a Danish physicist, who first proposed that energy levels of electrons are quantized and that excited electrons can only fall to discrete energy levels. This assignment illustrates the measurements that helped Bohr develop his original quantum model, as well as some practical uses for this science by measuring the emission spectra for mercury and sodium. Mercury vapor is used in fluorescent lights and sodium vapor is used in street lighting.

1. Start *Virtual ChemLab,* select *Atomic Theory,* and then select *Emission Spectra for Sodium and Mercury* from the list of assignments. The lab will open in the Quantum laboratory. A sample of gaseous sodium is on the lab bench in a sample tube and an alternating electric field of 300 V has been applied to cause the sodium gas to emit light. A spectrometer is on the right side of the lab bench and has been turned on. You can separate the light in an emission spectrum by using an optical prism or a diffraction grating. A spectrometer is an instrument designed to separate the emitted light into its component wavelengths. The detector window shows the output from the spectrometer.

2. Click on the ***Visible/Full*** switch in the detector window to change the output from the spectrometer to the visible spectrum. One spectral line is much more intense than all the others.

    What is the color and wavelength (in nm) of this line? (To determine the wavelength, move the cursor over the line and read the wavelength in the *x* field at the bottom of the detector window.)

    _____

Astronomers are excited about cities changing from normal incandescent streetlights to sodium vapor streetlights because astronomers can easily filter out the peak at 589 nm and minimize light pollution. Incandescent lights emit light at all wavelengths and make filtering impractical.

3. To exchange gas samples, click and drag the *Electric Field* and place it on the stockroom counter, and click and drag the *Gas (Na)* sample tube and place it on the stockroom counter as well. You may have to click on the main laboratory window in order to move the items.

4. Enter the stockroom by clicking in the *Stockroom*. Click on the *Gases* samples on the top shelf. Click on the cylinder labeled *Hg* to replace the Na in the sample tube with mercury vapor. If you point to the gas sample tube with the cursor it should read *Hg*.

5. Return to the laboratory and drag the gas sample tube off the stockroom counter and place it in the middle of the table as indicated by the spotlight. Drag the *Electric Field* and place it on the gas sample tube. Carefully click the button just above the left zero on the *Electric Field* controller and change the voltage to 300 V. Turn on the *Spectrometer* by clicking on the red/green button.

6. *How does the spectrum for mercury look different from sodium?* _____

    _____

Mercury vapor is used in the fluorescent light tubes that you see at school and home. The emitted light is not very bright for just the mercury vapor, but when scientists examined the full spectrum for mercury they saw what you just observed. There is an enormous emission in the ultraviolet region (UV). This light is sometimes called black light. You may have seen it with glow-in-the-dark displays.

Scientists coat the inside of the glass tube of fluorescent light tubes with a compound that will absorb UV and emit the energy as visible light with all the colors of the visible spectrum. All colors together create white light, which is why fluorescent light tubes emit very white light.

Laundry detergents contain compounds that absorb UV light and emit visible light. These compounds allow advertisers to claim *whiter and brighter whites and colors*. If you attend an event using black light, you may have seen your white socks or white shirt "glow."

2-1: Names and Formulas of Ionic Compounds

In this problem, you will go into the virtual laboratory and make a series of ionic compounds containing the cations Ag^+, Pb^+, Ca^{2+}, Fe^{3+}, and Cu^{2+}; observe the reactions and identify the color of the compound formed; write the chemical formulas; and write the chemical name.

1. Start *Virtual ChemLab,* select *Reactions and Stoichiometry,* and then select *Names and Formulas of Ionic Compounds* from the list of assignments. The lab will open in the Inorganic laboratory.

2. Drag a test tube from the box and place it on the metal test tube stand. You can then click on the bottle of Ag^+ ion solution on the shelf to add it to the test tube. Drag the test tube to the blue test tube rack.

3. Click on the **Divide** button on the edge of the lab bench four times to make four additional test tubes containing Ag^+. With one test tube in the metal stand and four others in the blue rack, click on the Na_2S bottle located on the lab bench. You will be able to observe what happens in the window at the top left. Record your observation in the table on the following page and write a correct chemical formula and name for the product of the reaction. If the solution remains clear, record NR, for no reaction. Drag this test tube to the red disposal bucket on the right.

4. Place a second tube from the blue rack (containing Ag^+) on the metal stand. Add Na_2SO_4. Record your observations and discard the tube. Use the next tube but add $NaCl$, and record your observations. Use the next tube but add $NaOH$, and record your observations. With the last tube add Na_2CO_3 and record your observations. When you are completely finished, click on the red disposal bucket to clear the lab.

5. Repeat steps 2-4 for Pb^{2+}, Ca^{2+}, Fe^{3+}, and Cu^{2+}. Complete the table on the following page.

Each cell should include a description of what you observed when the reagents were mixed and a correct chemical formula and name for all solutions that turned cloudy and NR for all solutions that did not react or remained clear. Remember to include roman numerals where appropriate.

	Ag^+	Pb^{2+}	Ca^{2+}	Fe^{3+}	Cu^{2+}
Na_2S (S^{2-})					
Na_2SO_4 (SO_4^{2-})					
$NaCl$ (Cl^-)					
$NaOH$ (OH^-)					
Na_2CO_3 (CO_3^{2-})					

2-2: Writing Balanced Precipitation Reactions

In this problem, you will go into the virtual laboratory and perform a series of precipitation reactions using Ag^+, Pb^{2+}, and Sb^{3+}. After observing the reactions, you will write the net ionic equations representing these reactions and then balance them.

1. Start *Virtual ChemLab,* select *Reactions and Stoichiometry,* and then select *Writing Balanced Precipitation Reactions* from the list of assignments. The lab will open in the Inorganic laboratory.

2. Drag a test tube from the box and place it on the metal test tube stand. You can then click on the bottle of *Ag⁺* ion solution (it is $AgNO_3$) on the shelf to add it to the test tube. Then drag the test tube to the blue test tube rack. Repeat for Pb^{2+} ($Pb(NO_3)_2$ and Sb^{3+} ($Sb(OH)_3$).

3. Moving your mouse over each test tube in the test-tube rack will allow you to identify it on the TV monitor. You can also click on the label at the top of each test tube in order to label the test tubes. Place the test tube containing the Ag⁺ solution in the metal test tube stand. You will be able to observe what happens in the window at the top left. Click on the *Na₂CO₃* reagent bottle to add it to the test tube in the stand.

   What color is the precipitate? Write a correct balanced net ionic equation for the reaction.

   _____

   When you have finished writing the equation and making your observations, place the test tube containing the precipitate in the red disposal bucket.

4. Place the test tube containing the Pb^{2+} solution in the metal test tube stand. Click on the *NaCl* reagent bottle to add it to the test tube in the stand.

   What color is the precipitate? Write a correct balanced net ionic equation for this reaction.

   _____

   When you have finished writing the equation and making your observations, place the test tube containing the precipitate in the red disposal bucket.

5. Place the test tube containing the Sb³⁺ solution in the metal test tube stand. Note that the antimony is not soluble and you will have a double displacement reaction. Click on the *Na₂S* reagent bottle to add it to the test tube in the stand.

   What color is the precipitate? Write a correct balanced net ionic equation for the reaction.

   _____

2-3: Strong and Weak Electrolytes

1. Start *Virtual ChemLab,* select *Reactions and Stoichiometry,* and then select *Strong and Weak Electrolytes* from the list of assignments. The lab will open in the Titration laboratory.

2. Enter the stockroom by clicking inside the *Stockroom* window. Once inside the stockroom, double-click or click and drag on the three reagents, *NaCl, Na₂CO₃* (100%), and *NaHCO₃* (100%) to move them to the stockroom counter. Click on the green *Return To Lab* arrow to return to the laboratory.

3. For each salt that was selected in the stockroom, complete the following procedure: double-click or click and drag the bottle on the stockroom counter to move it to the spotlight next to the balance. Click on the *Beakers* drawer and place a beaker in the spotlight next to the salt bottle in the balance area. Click on the *Balance* area to zoom in and open the bottle by clicking on the lid (*Remove Lid*). Drag a piece of weighing paper and drop it on the balance and then *Tare* the balance. Pick up the *Scoop* and scoop out some sample by first dragging the scoop to the mouth of the bottle and then pulling the scoop down the face of the bottle. As the scoop is dragged down the face of the bottle it will pick up different quantities of solid. Select the largest sample possible and drag the scoop to the balance and drop it on the weighing paper. This will put approximately 1 g of sample on the balance. Now drag the weighing paper with the sample and drop it in the beaker. Click on the green *Zoom Out* arrow to return to the laboratory.

   Move the beaker to the stir plate. Pick up the 25 mL graduated cylinder near the sink and hold it under the water tap until it fills. Pour the water into the beaker by dragging and dropping the cylinder on the beaker. Turn on the conductivity meter located on the lower right of the table and place the conductivity probe in the beaker and record the conductivity of the solution in the data table below. Double-click on the salt bottle to place it back on the stockroom counter. Place the beaker in the red disposal bucket. Repeat for the other two reagents.

4. When you have completed the three reagents, return to the *Stockroom*. Double-click on each bottle to return it to the shelf. Obtain three more samples (two salts and one solution): *KNO₃, NH₄Cl,* and *NH₃* and return to the laboratory. Follow procedure #3 for NH₄Cl and KNO₃, except in the case of NH₄Cl use only 0.6 g of sample (a 0.5 g scoop and a 0.1 g scoop).

   For the NH₃ solution, complete the following procedure: Place a beaker on the stir plate. Pick up the NH₃ solution from the stockroom shelf, drag it to the 25 mL graduated cylinder, and let go to fill the cylinder. The solution bottle will automatically go back to the stockroom shelf. Drag the 25 mL graduated cylinder to the beaker on the stir plate and drop it to transfer the solution into the beaker. Place the conductivity meter probe in the beaker and record the conductivity in the data table.

5. When you have completed the three reagents, return to the *Stockroom*. Double-click on each bottle to return it to the shelf. Obtain two more samples: *HCl* and *HCN.* Measure the conductivity of each solution following procedure #4 and record the conductivity in the data table.

Data Table

NaCl	Na₂CO₃	NaHCO₃	KNO₃
NH₄Cl	NH₃	HCl	HCN

6. Electrolytes are compounds that conduct electricity in aqueous solutions. *Which compounds in your table are electrolytes? Which are not electrolytes?* _____

_____

7. *Would any of these electrolytes conduct electricity in the solid form? Explain.* _____

_____

8. *Are these ionic or covalent compounds? Classify each compound in the grid as ionic or covalent. For a compound to be an electrolyte, what must happen when it dissolves in water?* _____

_____

_____

9. When an ionic solid dissolves in water, water molecules attract the ions causing them to dissociate or come apart. The resulting dissolved ions are electrically charged particles that allow the solution to conduct electricity. The following chemical equations represent this phenomenon:

$$NaCl\ (s)\ =\ Na^+\ (aq)\ +\ Cl^-\ (aq)$$
$$Na_2CO_3\ (s)\ =\ 2Na^+\ (aq)\ +\ CO_3^{2-}\ (aq)$$

Write a similar balanced chemical equation for each electrolyte in the data table.

10. *After examining the chemical reactions for the electrolytes, why does Na₂CO₃ have a higher conductivity than all of the other electrolytes?* _____

_____

_____

2-4: Precipitation Reactions

1. Start *Virtual ChemLab,* select *Reactions and Stoichiometry,* and then select *Precipitation Reactions* from the list of assignments. The lab will open in the Inorganic laboratory.

2. React each of the cations (across the top) with each of the anions (down the left) according to the data table using the following procedures:

Data Table

	AgNO$_3$ (Ag$^+$)	Pb(NO$_3$)$_2$ (Pb^{2+})	Ca(NO$_3$)$_2$ (Ca^{2+})
Na$_2$CO$_3$ (CO$_3^{2-}$)	a	f	k
Na$_2$S (S^{2-})	b	g	l
NaOH (OH$^-$)	c	h	m
Na$_2$SO$_4$ (SO$_4^{2-}$)	d	i	n
NaCl (Cl$^-$)	e	j	o

a. Drag a test tube from the box and place it on the metal test tube stand. You can then click on the bottle of Ag^+ ion solution on the shelf to add it to the test tube.

b. Click on the **Divide** button on the edge of the lab bench four times to make four additional test tubes containing Ag^+. With one test tube in the metal stand and four others in the blue rack, click on the Na$_2$CO$_3$ bottle on the reagent shelf and observe what happens in the window at the top left. Record your observation in the table above. If the solution remains clear, record NR for no reaction. Drag this test tube to the red disposal bucket on the right.

c. Drag a second tube from the blue rack to the metal stand. Add Na$_2$S, record your observations and discard the tube. Continue with the third, fourth and fifth tube, but add NaOH, Na$_2$SO$_4$, and NaCl respectively. Record your observations and discard the tubes. When you are finished, click on the red disposal bucket to clear the lab.

d. Repeat steps a-c for five test tubes of Pb^{2+} and Ca^{2+}. Record your observations in the data table. If no precipitate forms write NR for no reaction.

3. *What happens in grid space d? What other reactions give similar results? Is it necessary to write an*

    equation when no reaction occurs? Explain. _____

    _____

    _____

4. *Write balanced equations for all precipitation reactions you observed.*

    a.

    b.

    c.

    e.

    f.

    g.

    i.

    j.

    k.

    m.

5. *Write balanced net ionic equations for all precipitation reactions you observed.*

    a.

    b.

    c.

    e.

    f.

    g.

    i.

    j.

    k.

    m.

2-5: Counting Atoms

1. Start *Virtual ChemLab,* select *Reactions and Stoichiometry,* and then select *Counting Atoms(1)* from the list of assignments. The lab will open in the Calorimetry laboratory.

2. Enter the stockroom by clicking inside the *Stockroom* window. Once inside the stockroom, click on the *Metals* cabinet and then open the top drawer by clicking on it. When you open the drawer, a Petri dish will appear on the counter next to the cabinet. Place the sample of gold (Au) found in the drawer into the sample dish by double-clicking on the sample or by clicking and dragging it to the dish. Return to the stockroom view by clicking on the green *Zoom Out* arrow. Place the Petri dish on the stockroom counter by double-clicking on it or by clicking and dragging it to the counter. Click on the *Return to Lab* arrow to return to the laboratory.

3. Drag the Petri dish to the spotlight near the balance. Click on the *Balance* area to zoom in. Drag a piece of weighing paper to the balance pan, *Tare* the balance, and then drag the gold sample to the balance pan and record the mass.

   Mass =

4. *Calculate the moles of Au contained in the sample.*

5. *Calculate the atoms of Au contained in the sample.*

2-6: Counting Atoms

1. Start *Virtual ChemLab,* select *Reactions and Stoichiometry,* and then select *Counting Atoms (2)* from the list of assignments. The lab will open in the Calorimetry laboratory.

2. Enter the stockroom by clicking inside the *Stockroom* window. Once inside the stockroom, click on the *Metals* cabinet and then open the third drawer by clicking on it. When you open the drawer, a Petri dish will appear on the counter next to the cabinet. Place the sample of lead (Pb) found in the drawer into the sample dish by double-clicking on the sample or by clicking and dragging it to the dish. Return to the stockroom view by clicking on the green *Zoom Out* arrow. Place the Petri dish on the stockroom counter by double-clicking on it or by clicking and dragging it to the counter. Click on the *Return to Lab* arrow to return to the laboratory.

3. Drag the Petri dish to the spotlight near the balance. Click on the *Balance* area to zoom in. Drag a piece of weighing paper to the balance pan, **Tare** the balance, and then drag the lead sample to the balance pan and record the mass.

   Mass =

4. *Calculate the moles of Pb contained in the sample.*

5. *Calculate the atoms of Pb contained in the sample.*

6. Repeat steps 2-5 for uranium. Record the mass, the moles, and the atoms of uranium.

   Mass of uranium =
   Moles of uranium =
   Atoms of uranium =

2-7: Counting Atoms

1. Start *Virtual ChemLab,* select *Reactions and Stoichiometry,* and then select *Counting Atoms (3)* from the list of assignments. The lab will open in the Calorimetry laboratory.

2. Enter the stockroom by clicking inside the *Stockroom* window. Once inside the stockroom, click on the *Metals* cabinet and then open the first drawer by clicking on it. When you open the drawer, a Petri dish will appear on the counter next to the cabinet. Place the sample of erbium (Er) found in the drawer into the sample dish by double-clicking on the sample or by clicking and dragging it to the dish. Return to the stockroom view by clicking on the green *Zoom Out* arrow. Place the Petri dish on the stockroom counter by double-clicking on it or by clicking and dragging it to the counter. Click on the *Return to Lab* arrow to return to the laboratory.

3. Drag the Petri dish to the spotlight near the balance. Click on the *Balance* area to zoom in. Drag a piece of weighing paper to the balance pan, *Tare* the balance, and then drag the erbium sample to the balance pan and record the mass in the data and results table.

4. *Calculate the moles and the atoms of erbium and enter the results in the data and results table below.*

5. Repeat steps 2-4 for sodium, tungsten, and a metal of your choice.

Data and Results Table

	erbium (Er)	sodium (Na)	tungsten (W)	your choice
Mass (grams)				
Molar Mass (g/mol)				
Moles of each element				
Atoms of each element				

2-8: Counting Molecules

1. Start *Virtual ChemLab*, select *Reactions and Stoichiometry*, and then select *Counting Molecules (1)* from the list of assignments. The lab will open in the Calorimetry laboratory.

2. Enter the stockroom by clicking inside the *Stockroom* window. Once inside the stockroom, click on the sodium chloride (NaCl) bottle located on the Salts shelf and drag it to the stockroom counter. You can also double click on the bottle to move it to the counter, while the left and right blue arrow keys can be used to see additional bottles. Click on the *Return to Lab* arrow to return to the laboratory.

3. Drag the bottle to the spotlight near the balance, and click on the *Balance* area to zoom in. Drag a piece of weighing paper to the balance pan and then *Tare* the balance so the balance reads 0.0000 g. Click on the bottle lid (*Remove Lid*) to remove the lid.

4. Pick up the *Scoop* and scoop out some sample by first dragging the scoop to the mouth of the bottle and then pulling the scoop down the face of the bottle. As the scoop is dragged down the face of the bottle it will pickup different quantities of solid. Select the largest sample possible and drag the scoop to the weighing paper on the balance until it snaps in place and then let go. This will put approximately 1 g of sample on the balance.

   Record the mass of the sample. **Mass =** _____

5. *Calculate the moles of NaCl contained in the sample.*

6. *Calculate the moles of each element in NaCl.* _____

7. *Calculate the number of atoms for each element in NaCl.* _____

_____

2-9: Counting Molecules

1. Start *Virtual ChemLab,* select *Reactions and Stoichiometry,* and then select *Counting Molecules (2)* from the list of assignments. The lab will open in the Calorimetry laboratory.

2. Enter the stockroom by clicking inside the *Stockroom* window. Once inside the stockroom, click on the bottle containing table sugar (sucrose, $C_{12}H_{22}O_{11}$) located on the Organics shelf and drag it to the stockroom counter. You can also double click on the bottle to move it to the counter, while the left and right blue arrow keys can be used to see additional bottles. Click on the *Return to Lab* arrow to return to the laboratory.

3. Drag the bottle to the spotlight near the balance, and click on the *Balance* area to zoom in. Drag a piece of weighing paper to the balance pan and then *Tare* the balance so the balance reads 0.0000 g. Click on the bottle lid (*Remove Lid*) to remove the lid.

4. Pick up the *Scoop* and scoop out some sample by first dragging the scoop to the mouth of the bottle and then pulling the scoop down the face of the bottle. As the scoop is dragged down the face of the bottle it will pickup different quantities of solid. Select the largest sample possible and drag the scoop to the weighing paper on the balance until it snaps in place and then let go. This will put approximately 1 g of sample on the balance. Record the mass of the sample in the data and results table on the following page.

5. Repeat steps 2-4 for NH_4Cl (ammonium chloride) located on the Salts shelf and record the mass in the Data and Results Table.

6. *Calculate the moles of $C_{12}H_{22}O_{11}$ contained in the first sample and record your results in the data and results table.*

7. *Calculate the moles of each element in $C_{12}H_{22}O_{11}$ and record your results in the data and results table.*

8. *Calculate the number of atoms of each element in $C_{12}H_{22}O_{11}$ and record your results in the data and results table.*

9. Repeat steps 6-8 for NH_4Cl and record your results in the data and results table.

10. *Which of the compounds contains the most total number of atoms?* _____

Data and Results Table

	$C_{12}H_{22}O_{11}$	NH_4Cl
Mass (grams)		
Molar Mass (g/mol)		
Moles of compound		
Moles of each element		
Atoms of each element		

2-10: Counting Protons, Neutrons, and Electrons

1. Start *Virtual ChemLab,* select *Reactions and Stoichiometry,* and then select *Counting Protons, Neutrons, and Electrons (1)* from the list of assignments. The lab will open in the Calorimetry laboratory.

2. Enter the stockroom by clicking inside the *Stockroom* window. Once inside the stockroom, click on the *Metals* cabinet and then open the third drawer by clicking on it. When you open the drawer, a Petri dish will appear on the counter next to the cabinet. Place the sample of scandium (Sc) found in the drawer into the sample dish by double-clicking on the sample or by clicking and dragging it to the dish. Return to the stockroom view by clicking on the green *Zoom Out* arrow. Place the Petri dish on the stockroom counter by double-clicking on it or by clicking and dragging it to the counter. Click on the *Return to Lab* arrow to return to the laboratory.

3. Drag the Petri dish to the spotlight near the balance. Click on the *Balance* area to zoom in. Drag a piece of weighing paper to the balance pan, *Tare* the balance and drag the scandium sample to the balance pan.

    Record the mass of the sample. **Mass =** _____

4. *Calculate the moles of Sc contained in the sample.*

5. *Calculate the atoms of Sc contained in the sample.*

6. ^{45}Sc is the only naturally occurring isotope of scandium.

    How many protons, neutrons and electrons are there in one atom of ^{45}*Sc?* _____

    _____

7. *Calculate the number of protons, neutrons, and electrons in the sample of scandium that you weighed if it is 100% * ^{45}*Sc.*

2-11: Counting Protons, Neutrons, and Electrons

1. Start *Virtual ChemLab,* select *Reactions and Stoichiometry,* and then select *Counting Protons, Neutrons, and Electrons (2)* from the list of assignments. The lab will open in the Calorimetry laboratory.

2. Enter the stockroom by clicking inside the *Stockroom* window. Once inside the stockroom, click on the *Metals* cabinet and then open the top drawer by clicking on it. When you open the drawer, a Petri dish will appear on the counter next to the cabinet. Place the sample of bismuth (Bi) found in the drawer into the sample dish by double-clicking on the sample or by clicking and dragging it to the dish. Return to the stockroom view by clicking on the green *Zoom Out* arrow. Place the Petri dish on the stockroom counter by double-clicking on it or by clicking and dragging it to the counter. Click on the *Return to Lab* arrow to return to the laboratory.

3. Drag the Petri dish to the spotlight near the balance. Click on the *Balance* area to zoom in. Drag a piece of weighing paper to the balance pan, *Tare* the balance and drag the bismuth sample to the balance pan.

   Record the mass of the sample. *Mass =* _____

4. *Calculate the moles of Bi contained in the sample.*

5. *Calculate the atoms of Bi contained in the sample.*

6. ^{209}Bi is the only naturally occurring isotope of bismuth.

   How many protons, neutrons and electrons are there in one atom of ^{209}Bi*?* _____

   _____

7. *How many protons, neutrons and electrons are there in one ion of* $^{209}Bi^{5+}$*?* _____

   _____

8. *Calculate the number of protons, neutrons, and electrons in a sample of* $^{209}Bi^{5+}$ *that has the same mass as the bismuth sample you weighed.*

2-12: Creating a Solution of Known Molality

In this assignment, you will weigh out a sample of solid NH_4Cl and create a solution of known molality.

1. Start *Virtual ChemLab,* select *Reactions and Stoichiometry,* and then select *Creating a Solution of Known Molality* from the list of assignments. The lab will open in the Titration laboratory.

2. In the laboratory, a bottle of ammonium chloride (NH_4Cl) will be next to the balance and an empty beaker will be on the stir plate. Drag the beaker to the spotlight next to the balance, and click in the balance area to zoom in. Place the beaker on the balance and tare the balance. Click on the green *Zoom Out* arrow to return to the laboratory.

3. Drag the beaker to the sink and fill it with water until the beaker is approximately one-quarter full. Return the beaker to the balance and click in the *Balance* area to zoom in. Record the mass of the water in the data table. Drag the beaker off the balance to the spotlight on the right.

4. Place a weigh paper on the balance and tare the balance. Open the bottle by clicking on the lid (*Remove Lid*). Pick up the *Scoop* and scoop out some sample by first dragging the scoop to the mouth of the bottle and then pulling the scoop down the face of the bottle. As the scoop is dragged down the face of the bottle it will pickup different quantities of solid. Select the largest sample possible and drag the scoop to the weighing paper on the balance until it snaps in place and then let go. This will put approximately 1 g of sample on the balance. Repeat with a second *Scoop*. Record the mass of the NH_4Cl in the data table.

5. Drag the weigh paper to the beaker of water and add the NH_4Cl sample to the water to make an aqueous solution of NH_4Cl.

6. *Determine the kilograms of solvent (water) in the beaker and record the data in the data table.*

7. *Determine the moles of NH_4Cl in the sample and record the data in the data table.*

Data Table

mass NH_4Cl	
moles NH_4Cl	
mass water	
kg water	

8. The molality is calculated using the formula:

$$molality = \frac{\text{moles solute}}{\text{kg solvent}} = \frac{\text{mass solute}\big/\text{molar mass solute}}{\text{kg solvent}}$$

Calculate the molality of the NH₄Cl solution in units of mol/kg.

2-13: Creating a Solution of Known Molarity

In this assignment, you will weigh out a sample of baking soda ($NaHCO_3$) and create a solution of known molarity.

1. Start *Virtual ChemLab,* select *Reactions and Stoichiometry,* and then select *Creating a Solution of Known Molarity* from the list of assignments. The lab will open in the Titration laboratory.

2. In the laboratory, a bottle of baking soda (sodium bicarbonate, $NaHCO_3$) will be next to the balance, and an empty beaker will be on the stir plate. Drag the empty beaker to the spotlight next to the balance, click in the *Balance* area to zoom in, place a weigh paper on the balance, and tare the balance.

3. Open the bottle by clicking on the lid (*Remove Lid*). Pick up the *Scoop* and scoop out some sample by first dragging the scoop to the mouth of the bottle and then pulling the scoop down the face of the bottle. As the scoop is dragged down the face of the bottle it will pickup different quantities of solid. Select the largest sample possible and drag the scoop to the weighing paper on the balance until it snaps in place and then let go. This will put approximately 1 g of sample on the balance. Repeat with a second *Scoop*. Record the mass of the $NaHCO_3$ in the data table.

4. Drag the weigh paper to the beaker and add the $NaHCO_3$ sample to the beaker. Click on the green *Zoom Out* arrow to return to the laboratory.

5. Drag the beaker to the 50 mL graduated cylinder (the largest one) by the sink and empty the sample into the cylinder. Hold the cylinder under the tap until it fills with water to make an aqueous solution of $NaHCO_3$. (When the graduated cylinder is full it will automatically snap back into place.) Note that the solid is added and dissolved before the volume is measured when making a molar solution. Chemists normally use a volumetric flask for making molar solutions, but this is not available in the simulation.

6. *Determine the liters of solution in the cylinder and record the data in the data table.*

7. *Determine the moles of NaHCO₃ in the sample and record the data in the data table.*

Data Table

mass $NaHCO_3$	
moles $NaHCO_3$	
liters solution	

8. The molarity is calculated using the formula:

$$molarity = \frac{moles\ solute}{L\ solution} = \frac{mass\ solute/molar\ mass\ solute}{L\ solution}$$

Calculate the molarity of the NaHCO₃ solution in units of mol/L.

2-14: Converting Concentrations to Different Units

Occasionally, when making solutions in the laboratory, it is convenient to make a solution with a certain concentration unit, such as molarity, and then convert the concentration to a different unit. In this assignment, you will make a sodium bicarbonate (baking soda) solution of a certain molarity and then convert that concentration to molality, mass percent, and mole fraction.

1. Start *Virtual ChemLab,* select *Reactions and Stoichiometry,* and then select *Converting Concentrations to Different Units* from the list of assignments. The lab will open in the Titration laboratory.

2. In the laboratory, a bottle of baking soda (sodium bicarbonate, $NaHCO_3$) will be next to the balance, and an empty beaker will be on the stir plate. Drag the empty beaker to the spotlight next to the balance, click in the *Balance* area to zoom in, place a weigh paper on the balance, and tare the balance.

3. Open the bottle by clicking on the lid (*Remove Lid*). Pick up the *Scoop* and scoop out some sample by first dragging the scoop to the mouth of the bottle and then pulling the scoop down the face of the bottle. As the scoop is dragged down the face of the bottle, it will pick up different quantities of solid. Select the largest sample possible and drag the scoop to the weighing paper on the balance until it snaps in place and then let go. This will put approximately 1 g of sample on the balance. Repeat this process six additional times so there is approximately 7.0 g of sample. Record the mass of the $NaHCO_3$ in the data table.

4. Drag the weigh paper to the beaker and add the $NaHCO_3$ sample to the beaker. Click on the green *Zoom Out* arrow to return to the laboratory.

5. Drag the beaker to the 50 mL graduated cylinder (the largest one) by the sink and empty the sample into the cylinder. Hold the cylinder under the tap until it fills with water to make an aqueous solution of $NaHCO_3$. (When the graduated cylinder is full, it will automatically snap back into place.) Note that the solid is added and dissolved before the volume is measured when making a molar solution. Chemists normally use a volumetric flask for making molar solutions, but this is not available in the simulation. Record the volume of the solution, in L, in the data table.

6. *Calculate the moles of $NaHCO_3$ in the sample and record the data in the data table.*

Data Table

mass $NaHCO_3$	
moles $NaHCO_3$	
liters $NaHCO_3$	

7. The molarity is calculated using the formula:

$$\text{molarity} = \frac{\text{moles solute}}{\text{L solution}} = \frac{\text{mass solute}/\text{molar mass solute}}{\text{L solution}}$$

Calculate the molarity of the NaHCO₃ solution in units of mol/L.

8. *If the density of the solution is 1.047 g/mL, calculate the molality of the solution in units of mol/kg.*

9. *Calculate the mass percent of sodium bicarbonate in the solution.*

10. *Calculate the mole fraction of sodium bicarbonate in the solution.*

3-1: Endothermic vs. Exothermic

In various chemical processes such as reactions and the dissolving of salts, heat is either absorbed or given off. We call these events either an endothermic (heat in) or exothermic (heat out) process. It is usual to detect these heat events by measuring the temperature change associated with the process. In this problem, you will dissolve several salts in water, measure the resulting temperature change, and then make deductions about the nature of the process.

1. Start *Virtual ChemLab,* select *Thermodynamics,* and then select *Endothermic vs. Exothermic* from the list of assignments. The lab will open in the Calorimetry laboratory.

2. There will be a bottle of sodium chloride (NaCl) on the lab bench. A weigh paper will be on the balance with approximately 2 g of NaCl on the paper.

3. The calorimeter will be on the lab bench and filled with 100 mL water. Click the *Lab Book* to open it. Make certain the stirrer is **On** (you should be able to see the shaft rotating). In the thermometer window click **Save** to begin recording data. Allow 20-30 seconds to obtain a baseline temperature of the water.

4. Drag the weigh paper with the sample to the calorimeter until it snaps into place and then pour the sample into the calorimeter. Observe the change in temperature until it reaches a maximum and then record data for an additional 20-30 seconds. Click **Stop**. (You can click on the clock on the wall labeled *Accelerate* to accelerate the time in the laboratory.) A data link icon will appear in the lab book. Click the data link icon and record the temperature before adding the NaCl and the *highest* or *lowest* temperature after adding the NaCl in the data table.

5. Click the red disposal bucket to clear the lab. Click on the Stockroom to enter. Click on the clipboard and select the preset experiment called *Heat of Solution-NaNO₃* and repeat the experiment with NaNO₃. Record the initial and final temperatures in the data table.

6. Click the red disposal bucket to clear the lab. Click on the Stockroom to enter. Click the clipboard and select the preset experiment called *Heat of Solution-NaAc* and repeat the experiment with $NaCH_3COO$ (NaAc). Record the initial and final temperatures in the data table.

Data Table

Mixture	T_1	T_2	$\Delta T\ (T_2 T_1)$
NaCl (s) + H_2O (l)			
$NaNO_3$ (s) + H_2O (l)			
$NaCH_3COO$ + H_2O (l)			

Use your experimental data to answer the following questions.

7. *Calculate ΔT ($\Delta T = T_2 - T_1$) for each mixture and record it in the data table.*

8. An exothermic process gives off heat (warms up). An endothermic process absorbs heat (cools off).

    Which solutions are endothermic and which are exothermic? What is the sign of the change in

    enthalpy ()H) in each case? _____

    _____

9. *Which solution(s) had little or no change in temperature?* _____

3-2: Enthalpy of Solution: NH₄NO₃

Have you ever used one of those "instant cold packs" that looks like a plastic bag filled with liquid? If you hit the bag and shake it up it gets extremely cold, but why does it do that? The liquid inside the cold pack is water, and in the water is another plastic bag or tube containing NH_4NO_3 fertilizer. When you hit the cold pack, it breaks the tube so that the water mixes with the fertilizer. The dissolving of a salt, such as NH_4NO_3, in water is called dissolution, and the heat associated with the dissolving process is called the Enthalpy of Solution. In this problem, you will take a sample of NH_4NO_3, dissolve it in water, and after measuring the change in temperature, calculate the enthalpy of solution for NH_4NO_3.

1. Start *Virtual ChemLab,* select *Thermodynamics,* and then select *Enthalpy of Solution: NH₄NO₃* from the list of assignments. The lab will open in the Calorimetry laboratory.

2. There will be a bottle of ammonium nitrate (NH_4NO_3) on the lab bench. A weigh paper will be on the balance with approximately 2 g of NH_4NO_3 on the paper. Record the mass of the sample in the data table. If you cannot read the mass on the balance, click in the balance area to *Zoom In. Return to Lab* when you have recorded the mass.

3. The coffee cup calorimeter will be on the lab bench and filled with 100 mL water. Click the *Lab Book* to open it. Make certain the stirrer is *On* (you should be able to see the shaft rotating). In the thermometer window click *Save* to begin recording data. Allow 20-30 seconds to obtain a baseline temperature of the water.

4. Drag the weigh paper with the sample to the calorimeter until it snaps into place and then pour the sample into the calorimeter. Observe the change in temperature until it reaches a maximum and then record data for an additional 20-30 seconds. Click *Stop*. (You can click on the clock on the wall labeled *Accelerate* to accelerate the time in the laboratory.) A data link icon will appear in the lab book. Click the data link icon and record the temperature before adding the NH_4NO_3 and the *highest* or *lowest* temperature after adding the NH_4NO_3 in the data table.

Data Table

Mixture	mass	$T_{initial}$	T_{final}
NH_4NO_3 (s) + H_2O (l)			

5. *Calculate ΔT ($\Delta T = T_{initial} - T_{final}$) for the dissolving process.* _____

6. An exothermic process gives off heat (warms up), and an endothermic process absorbs heat (cools off).

   Was the addition of NH₄NO₃ to the water an endothermic or exothermic process? What is the sign of

   the change in enthalpy ()H)? _____

7. *Determine the moles of NH₄NO₃ in the sample.* The molecular weight of NH_4NO_3 is 80 g/mol.

8. The heat absorbed or lost by the water can be calculated using $q = m \cdot C_{water} \cdot \Delta T$. Assume that the density of water is 1 g/mL.

Calculate the mass of the water and substitute for m. ΔT is the change in the temperature of the water and C_{water} is the specific heat capacity for water (4.184 J/g·K). What is the heat absorbed or lost, in J,

by the water? _____

9. The heat transferred from/to the NH_4NO_3 can be divided by the moles of NH_4NO_3 to obtain the molar heat of solution for NH_4NO_3.

    What is the molar heat of solution, in kJ, of NH_4NO_3? _____

    _____

10. *If the accepted value for the heat of solution for sugar is 25.69 kJ/mol, calculate the percent error.*

$$\% \ Error = \frac{|your \ answer - accepted \ answer|}{accepted \ answer} \times 100$$

% Error =

This experiment does not consider that all of the conditions are standard state conditions; therefore, you are calculating ΔH_{sol} **not** $\Delta H°_{sol}$.

3-3: Specific Heat of Al

On a sunny day, the water in a swimming pool may warm up a degree or two while the concrete around the pool may become too hot to walk on with bare feet. This may seem strange since the water and concrete are being heated by the same source—the sun. This evidence suggests that it takes more heat to raise the temperature of some substances than others, which is true. The amount of heat required to raise the temperature of 1 g of a substance by 1 degree is called the *specific heat capacity* or *specific heat* of that substance. Water, for instance, has a specific heat of 4.18 J/K·g. This value is high in comparison with the specific heats for other materials, such as concrete or metals. In this experiment, you will use a simple calorimeter and your knowledge of the specific heat of water to measure the specific heat of aluminum (Al).

1. Start *Virtual ChemLab*, select *Thermodynamics,* and then select *Specific Heat of Al* from the list of assignments. The lab will open in the Calorimetry laboratory.

2. Click on the *Lab Book* to open it. Record the mass of Al on the balance. If it is too small to read click on the *Balance* area to zoom in, record the mass of Al in the data table below, and return to the laboratory.

3. Pick up the Al sample from the balance pan and place the sample in the oven. Click the oven door to close. The oven is set to heat to 200°C.

4. The calorimeter has been filled with 100 mL water. The density of water at 25°C is 0.998 g/mL. Use the density of the water to determine the mass of water from the volume and record the volume and mass in the data table.

    Make certain the stirrer is **On** (you should be able to see the shaft rotating). In the thermometer window, click **Save** to begin recording data. Allow 20-30 seconds to obtain a baseline temperature of the water. You can observe the temperature in the calorimeter as a function of time using the graph window.

5. Click on the *Oven* to open it. Drag the hot Al sample from the oven until it snaps into place above the calorimeter and drop it in. Click the thermometer and graph windows to bring them to the front and observe the change in temperature in the graph window until it reaches a constant value and then wait an additional 20-30 seconds. Click **Stop** in the temperature window. (You can click on the clock on the wall labeled *Accelerate* to accelerate the time in the laboratory.) A data link icon will appear in the lab book. Click the data link icon and record the temperature before adding the Al and the *highest* temperature after adding the Al in the data table. (Remember that the water will begin to cool down after reaching the equilibrium temperature.)

Data Table

	Al
mass of metal (g)	
volume of water (mL)	
mass of water (g)	
initial temperature of water (°C)	
initial temperature of metal (°C)	
max temp of water + metal (°C)	

6. *Calculate the change in temperature of the water (ΔT_{water}).* _____

_____

7. *Calculate the heat (q), in J, gained by the water using the following equation:*

$$q_{water} = m_{water} \times \Delta T_{water} \times C_{water} \text{, given } C_{water} = 4.184 \text{ J/(K·g)}$$

_____

8. *Calculate the changes in temperature of the Al (ΔT_{Al}).* _____

_____

9. *Remembering that the heat gained by the water is equal to the heat lost by the metal, calculate the specific heat of aluminum in J/K≅g.*

$$q_{water} = -q_{metal} = m_{Al} \times \Delta T_{Al} \times C_{Al} \text{ and } C_{Al} = \frac{q_{metal}}{(m_{metal})(\Delta T_{metal})}$$

_____

10. *Calculate the percent error in the specific heat value that you determined experimentally.* The accepted value for Al is 0.903 J/ K·g.

$$\% \, Error = \frac{|your \; answer - accepted \; answer|}{accepted \; answer} \times 100$$

$$\% \, Error =$$

3-4: Specific Heat of Pb

On a sunny day, the water in a swimming pool may warm up a degree or two while the concrete around the pool may become too hot to walk on with bare feet. This may seem strange since the water and concrete are being heated by the same source—the sun. This evidence suggests that it takes more heat to raise the temperature of some substances than others, which is true. The amount of heat required to raise the temperature of 1 g of a substance by 1 degree is called the *specific heat capacity* or *specific heat* of that substance. Water, for instance, has a specific heat of 4.18 J/K·g. This value is high in comparison with the specific heats for other materials, such as concrete or metals. In this experiment, you will use a simple calorimeter and your knowledge of the specific heat of water to measure the specific heat of lead (Pb).

1. Start *Virtual ChemLab,* select *Thermodynamics,* and then select *Specific Heat of Pb* from the list of assignments. The lab will open in the Calorimetry laboratory.

2. Click on the *Lab Book* to open it. Record the mass of Pb on the balance. If it is too small to read click on the *Balance* area to zoom in, record the mass of Pb in the data table below, and return to the laboratory.

3. Pick up the Pb sample from the balance pan and place the sample in the oven. Click the oven door to close. The oven is set to heat to 200°C.

4. The calorimeter has been filled with 100 mL water. The density of water at 25°C is 0.998 g/mL. Use the density of the water to determine the mass of water from the volume and record the volume and mass in the data table.

   Make certain the stirrer is **On** (you should be able to see the shaft rotating). In the thermometer window, click **Save** to begin recording data. Allow 20-30 seconds to obtain a baseline temperature of the water. You can observe the temperature in the calorimeter as a function of time using the graph window.

5. Click on the *Oven* to open it. Drag the hot Pb sample from the oven until it snaps into place above the calorimeter and drop it in. Click the thermometer and graph windows to bring them to the front and observe the change in temperature in the graph window until it reaches a constant value and then wait an additional 20-30 seconds. Click **Stop** in the temperature window. (You can click on the clock on the wall labeled *Accelerate* to accelerate the time in the laboratory.) A data link icon will appear in the lab book. Click the data link icon and record the temperature before adding the Pb and the *highest* temperature after adding the Pb in the data table. (Remember that the water will begin to cool down after reaching the equilibrium temperature.)

Data Table

	Pb
mass of metal (g)	
volume of water (mL)	
mass of water (g)	
initial temperature of water (°C)	
initial temperature of metal (°C)	
max temp of water + metal (°C)	

6. *Calculate the changes in temperature of the water (ΔT_{water}).* _____

    _____

7. *Calculate the heat (q), in J, gained by the water using the following equation:*

$$q_{water} = m_{water} \times \Delta T_{water} \times C_{water} \text{, given } C_{water} = 4.184 \text{ J/(K·g)}$$

    _____

8. *Calculate the changes in temperature of the Pb (ΔT_{Pb}).* _____

    _____

9. *Remembering that the heat gained by the water is equal to the heat lost by the metal, calculate the specific heat of lead in J/K≅g.*

$$q_{water} = -q_{metal} = m_{Pb} \times \Delta T_{Pb} \times C_{Pb} \text{ and } C_{Pb} = \frac{q_{metal}}{(m_{metal})(\Delta T_{metal})}$$

    _____

10. *Calculate the percent error in the specific heat value that you determined experimentally.* The accepted value for Pb is 0.130 J/ K·g.

$$\% \text{ Error} = \frac{|your\ answer - accepted\ answer|}{accepted\ answer} \times 100$$

    $\% \text{ Error} =$

3-5: Heat of Combustion: Chicken Fat

The heat of combustion (ΔH_{comb}) is the heat of reaction for the complete burning (reacting with O_2) of one mole of a substance to form CO_2 and H_2O. Calorimetry experiments that measure the heat of combustion can be performed at constant volume using a device called a bomb calorimeter. In a bomb calorimeter a sample is burned in a constant-volume chamber in the presence of oxygen at high pressure. The heat that is released warms the water surrounding the chamber. By measuring the temperature increase of the water, it is possible to calculate the quantity of heat released during the combustion reaction. In this assignment you will calculate the heat of combustion of chicken fat. The calorimeter has already been calibrated by combusting benzoic acid.

1. Start *Virtual ChemLab*, select *Thermodynamics*, and then select *Heat of Combustion: Chicken Fat* from the list of assignments. The lab will open in the Calorimetry laboratory with the bomb calorimeter out and disassembled and with a sample of chicken fat in the calorimeter cup on the balance. The balance has already been tared.

2. Click on the *Lab Book* to open it.

3. Record the mass of the chicken fat sample from the balance. If you cannot read it click on the *Balance* area to zoom in, record the mass in the data table below and return to the laboratory.

4. Double-click the following (in numerical order) to assemble the calorimeter: (1) the cup on the balance pan, (2) the bomb head, (3) the screw cap, and (4) the bomb. Click the calorimeter lid to close it. Combustion experiments can take a considerable length of time. Click the clock on the wall labeled *Accelerate* to accelerate the laboratory time.

5. Click the bomb control panel and the plot window to bring them to the front. Click on the *Save* button to save data to the lab book. Allow the graph to proceed for 20-30 seconds to establish a baseline temperature.

6. Click *Ignite* and observe the graph. When the temperature has leveled off (up to 5 minutes of laboratory time), click *Stop*. A data link icon will appear in the lab book. Click the data link icon to view the collected data. Record the temperature before and after ignition of the chicken fat sample in the data table.

Data Table

	chicken fat
mass of sample (g)	
initial temperature (°C)	
final temperature (°C)	

7. *Calculate ΔT for the water using $\Delta T = |T_f - T_i|$.* _____

8. *Calculate the moles of chicken fat in the sample (MW_{fat} = 797.7 g/mol).*

9. ΔH_{comb} for chicken fat can be calculated using $\Delta H_{comb} = (C_{system}\Delta T)/n$, where n is the moles of chicken fat in the sample and C_{system} is the heat capacity of the calorimetric system.

    Use 10.310 kJ/K for C_{system} and calculate the heat of combustion, in kJ/mol, for chicken fat.

10. *If the accepted value for the heat of combustion for chicken fat is 30,038 kJ/mol calculate the percent error.*

$$\% \, Error = \frac{|your\ answer - accepted\ answer|}{accepted\ answer} \times 100$$

    % Error =

    This experiment does not consider that all of the conditions are standard state conditions; therefore, you are calculating ΔH_{comb} **not** ΔH°_{comb}.

11. The "calorie" used to measure the caloric content of foods is actually a kilocalorie (kcal) or 4184 kJ.

    If the heat of combustion for sugar is 5639 kJ/mol, why are people who are on limited calorie diets

    advised to limit their fat intake? _____

    _____

    _____

12. The food that we ingest is certainly not "combusted" in the same manner as is done in a bomb calorimeter.

    Why can we compare the heats of combustion of sugar or chicken fat measured in a bomb

    calorimeter with the caloric content of those foods? _____

    _____

    _____

    _____

3-6: Heat of Combustion: Sugar

The heat of combustion (ΔH_{comb}) is the heat of reaction for the complete burning (reacting with O_2) of one mole of a substance to form CO_2 and H_2O. Calorimetry experiments that measure the heat of combustion can be performed at constant volume using a device called a bomb calorimeter. In a bomb calorimeter a sample is burned in a constant-volume chamber in the presence of oxygen at high pressure. The heat that is released warms the water surrounding the chamber. By measuring the temperature increase of the water, it is possible to calculate the quantity of heat released during the combustion reaction. In this assignment you will calculate the heat of combustion of sugar (sucrose, $C_{12}H_{22}O_{11}$). The calorimeter has already been calibrated by combusting benzoic acid.

1. Start *Virtual ChemLab,* select *Thermodynamics,* and then select *Heat of Combustion: Sugar* from the list of assignments. The lab will open in the Calorimetry laboratory with the bomb calorimeter out and disassembled and with a sample of sugar in the calorimeter cup on the balance. The balance has already been tared.

2. Click on the *Lab Book* to open it.

3. Record the mass of the sugar sample from the balance. If you cannot read it click on the *Balance* area to zoom in, record the mass in the data table below and return to the laboratory.

4. Double-click the following (in numerical order) to assemble the calorimeter: (1) the cup on the balance pan, (2) the bomb head, (3) the screw cap, and (4) the bomb. Click the calorimeter lid to close it. Combustion experiments can take a considerable length of time. Click the clock on the wall labeled *Accelerate* to accelerate the laboratory time.

5. Click the bomb control panel and the plot window to bring them to the front. Click on the **Save** button to save data to the lab book. Allow the graph to proceed for 20-30 seconds to establish a baseline temperature.

6. Click **Ignite** and observe the graph. When the temperature has leveled off (up to 5 minutes of laboratory time), click **Stop**. A data link icon will appear in the lab book. Click the data link icon to view the collected data. Record the temperature before and after ignition of the sugar sample in the data table.

Data Table

	sucrose ($C_{12}H_{22}O_{11}$)
mass of sample (g)	
initial temperature (°C)	
final temperature (°C)	

7. *Write a complete balanced chemical equation for the combustion of sucrose.* _____

_____

8. *Calculate ΔT for the water using $\Delta T = |T_f - T_i|$.* _____

9. *Calculate the moles of sucrose in the sample (MW$_{sucrose}$ = 342.3 g/mol).*

10. ΔH_{comb} for sucrose can be calculated using $\Delta H_{comb} = \left(C_{system}\Delta T\right)/n$, where n is the moles of sucrose in the sample and C_{system} is the heat capacity of the calorimetric system.

    Use 10.310 kJ/K for C$_{system}$ and calculate the heat of combustion, in kJ/mol, for sucrose.

11. *If the accepted value for the heat of combustion for sugar is 5639 kJ/mol calculate the percent error.*

$$\% \; Error = \frac{\left|your \; answer - accepted \; answer\right|}{accepted \; answer} \times 100$$

% Error =

This experiment does not consider that all of the conditions are standard state conditions; therefore, you are calculating ΔH_{comb} **not** ΔH°_{comb}.

3-7: Heat of Combustion: TNT

The heat of combustion (ΔH_{comb}) is the heat of reaction for the complete burning (reacting with O_2) of one mole of a substance to form CO_2 and H_2O. Calorimetry experiments that measure the heat of combustion can be performed at constant volume using a device called a bomb calorimeter. In a bomb calorimeter a sample is burned in a constant-volume chamber in the presence of oxygen at high pressure. The heat that is released warms the water surrounding the chamber. By measuring the temperature increase of the water, it is possible to calculate the quantity of heat released during the combustion reaction. In this assignment you will calculate the heat of combustion of 2,4,6-trinitrotolune (TNT). The calorimeter has already been calibrated by combusting benzoic acid.

1. Start *Virtual ChemLab,* select *Thermodynamics,* and then select *Heat of Combustion: TNT* from the list of assignments. The lab will open in the Calorimetry laboratory with the bomb calorimeter out and disassembled and with a sample of TNT in the calorimeter cup on the balance. The balance has already been tared.

2. Click on the *Lab Book* to open it.

3. Record the mass of the TNT sample from the balance. If you cannot read it click on the *Balance* area to zoom in, record the mass in the data table below and return to the laboratory.

4. Double-click the following (in numerical order) to assemble the calorimeter: (1) the cup on the balance pan, (2) the bomb head, (3) the screw cap, and (4) the bomb. Click the calorimeter lid to close it. Combustion experiments can take a considerable length of time. Click the clock on the wall labeled *Accelerate* to accelerate the laboratory time.

5. Click the bomb control panel and the plot window to bring them to the front. Click on the ***Save*** button to save data to the lab book. Allow the graph to proceed for 20-30 seconds to establish a baseline temperature.

6. Click ***Ignite*** and observe the graph. When the temperature has leveled off (up to 5 minutes of laboratory time), click ***Stop***. A data link icon will appear in the lab book. Click the data link icon to view the collected data. Record the temperature before and after ignition of the TNT sample in the data table.

Data Table

	2,4,6-trinitrotoluene (TNT)
mass of sample (g)	
initial temperature (°C)	
final temperature (°C)	

7. *Calculate ΔT for the water using $\Delta T = |T_f - T_i|$.* _____

8. *Calculate the moles of TNT in the sample ($MW_{TNT} = 227.13$ g/mol).*

9. ΔH_{comb} for TNT can be calculated using $\Delta H_{comb} = \left(C_{system} \Delta T \right) / n$, where n is the moles of TNT in the sample and C_{system} is the heat capacity, in kJ/mol, of the calorimetric system.

   Use 10.310 kJ/K for C_{system} and calculate the heat of combustion for TNT.

10. *If the accepted value for the heat of combustion for TNT is 3406 kJ/mol, calculate the percent error.*

$$\% \; Error = \frac{\left| your \; answer - accepted \; answer \right|}{accepted \; answer} \times 100$$

   % Error =

   This experiment does not consider that all of the conditions are standard state conditions; therefore, you are calculating ΔH_{comb} **not** ΔH°_{comb}.

11. *The heat of combustion for sugar is 5639 kJ/mole, but that for TNT is 3406 kJ/mole. Why, if the heat*

   of combustion for TNT is smaller than for sugar, is TNT an explosive? _____

   _____

   _____

3-8: Heat of Formation: Ethanol

The heat of formation is the heat of reaction for the formation of a compound from its elements. The heat of formation can be determined by measuring the heat of combustion for the compound and then using Hess's law to convert the heat of combustion to a heat of formation. Calorimetry experiments that measure the heat of combustion can be performed at constant volume using a device called a bomb calorimeter. In a bomb calorimeter a sample is burned in a constant-volume chamber in the presence of oxygen at high pressure. The heat that is released warms the water surrounding the chamber. By measuring the temperature increase of the water, it is possible to calculate the quantity of heat released during the combustion reaction. In this assignment you will first measure the heat of combustion of ethanol (ethyl alcohol, C_2H_5OH) and then convert the heat of combustion to a heat of formation.

1. Start *Virtual ChemLab,* select *Thermodynamics,* and then select *Heat of Formation: Ethanol* from the list of assignments. The lab will open in the Calorimetry laboratory with the bomb calorimeter out and disassembled and with a sample of ethanol in the calorimeter cup on the balance. The balance has already been tared.

2. Click on the *Lab Book* to open it.

3. Record the mass of the ethanol sample from the balance. If you cannot read it click on the *Balance* area to zoom in, record the mass in the data table below and return to the laboratory.

4. Double-click the following (in numerical order) to assemble the calorimeter: (1) the cup on the balance pan, (2) the bomb head, (3) the screw cap, and (4) the bomb. Click the calorimeter lid to close it. Combustion experiments can take a considerable length of time. Click the clock on the wall labeled *Accelerate* to accelerate the laboratory time.

5. Click the bomb control panel and the plot window to bring them to the front. Click on the *Save* button to save data to the lab book. Allow the graph to proceed for 20-30 seconds to establish a baseline temperature.

6. Click *Ignite* and observe the graph. When the temperature has leveled off (up to 5 minutes of laboratory time), click *Stop*. A data link icon will appear in the lab book. Click the data link icon to view the collected data. Record the temperature before and after ignition of the ethanol sample in the data table.

Data Table

	ethanol (C_2H_5OH)
mass of sample (g)	
initial temperature (°C)	
final temperature (°C)	

7. *Write a complete balanced chemical equation for the combustion of ethanol.*

_____

8. *Calculate ΔT for the water using $\Delta T = |T_f - T_i|$.* _____

9. *Calculate the moles of ethanol in the sample (MW$_{enthanol}$ = 46.00 g/mol).*

10. ΔH_{comb} for ethanol can be calculated using $\Delta H_{comb} = (C_{system}\Delta T)/n$, where n is the moles of ethanol in the sample and C_{system} is the heat capacity of the calorimetric system.

    Use 10.310 kJ/K for C_{system} and calculate the heat of combustion, in kJ/mol, for ethanol. The heat of combustion will be negative since it is an exothermic reaction.

11. *Write an equation for the combustion of ethanol in the form $\Delta H_{comb} = \Sigma\, n\Delta H_f\ (products) - \Sigma\, m\Delta H_f$*

    (reactants). _____

12. *Calculate the heat of formation for C_2H_5OH, given the standard enthalpies of formation for CO_2, H_2O, and O_2 are -393.5 kJ/mol, -285.83 kJ/mol, and 0 kJ/mol, respectively.*

13. *If the accepted value for the enthalpy of formation for ethanol is −277.7 kJ/mol, calculate the percent error.*

    $$\% \ Error = \frac{|\ your\ answer - accepted\ answer\ |}{accepted\ answer} \times 100$$

    % Error =

    This experiment does not consider that all of the conditions are standard state conditions; therefore, you are calculating ΔH_f **not** ΔH°_f.

3-9: Heat of Formation: Aspirin

The heat of formation is the heat of reaction for the formation of a compound from its elements. The heat of formation can be determined by measuring the heat of combustion for the compound and then using Hess's law to convert the heat of combustion to a heat of formation. Calorimetry experiments that measure the heat of combustion can be performed at constant volume using a device called a bomb calorimeter. In a bomb calorimeter a sample is burned in a constant-volume chamber in the presence of oxygen at high pressure. The heat that is released warms the water surrounding the chamber. By measuring the temperature increase of the water, it is possible to calculate the quantity of heat released during the combustion reaction. In this assignment you will first measure the heat of combustion of aspirin ($C_9H_8O_4$) and then convert the heat of combustion to a heat of formation.

1. Start *Virtual ChemLab*, select *Thermodynamics*, and then select *Heat of Formation: Aspirin* from the list of assignments. The lab will open in the Calorimetry laboratory with the bomb calorimeter out and disassembled and with a sample of asprin in the calorimeter cup on the balance. The balance has already been tared.

2. Click on the *Lab Book* to open it.

3. Record the mass of the aspirin sample from the balance. If you cannot read it click on the *Balance* area to zoom in, record the mass in the data table below and return to the laboratory.

4. Double-click the following (in numerical order) to assemble the calorimeter: (1) the cup on the balance pan, (2) the bomb head, (3) the screw cap, and (4) the bomb. Click the calorimeter lid to close it. Combustion experiments can take a considerable length of time. Click the clock on the wall labeled *Accelerate* to accelerate the laboratory time.

5. Click the bomb control panel and the plot window to bring them to the front. Click on the *Save* button to save data to the lab book. Allow the graph to proceed for 20-30 seconds to establish a baseline temperature.

6. Click *Ignite* and observe the graph. When the temperature has leveled off (up to 5 minutes of laboratory time), click *Stop*. A data link icon will appear in the lab book. Click the data link icon to view the collected data. Record the temperature before and after ignition of the aspirin sample in the data table.

Data Table

	aspirin ($C_9H_8O_4$)
mass of sample (g)	
initial temperature (°C)	
final temperature (°C)	

7. *Write a complete balanced chemical equation for the combustion of aspirin.* _____

_____

8. *Calculate ΔT for the water using $\Delta T = |T_f - T_i|$.* _____

9. *Calculate the moles of aspirin in the sample ($MW_{aspirin}$ = 180.00 g/mol).*

10. ΔH_{comb} for aspirin can be calculated using $\Delta H_{comb} = \left(C_{system} \Delta T \right)/n$, where n is the moles of aspirin in the sample and C_{system} is the heat capacity of the calorimetric system.

    Use 10.310 kJ/K for C_{system} and calculate the heat of combustion, in kJ/mol, for aspirin. The heat of combustion will be negative since it is an exothermic reaction.

11. *Write an equation for the combustion of aspirin in the form $\Delta H_{comb} = \Sigma n \Delta H_f$ (products) $- \Sigma m \Delta H_f$ (reactants).*

12. *Calculate the heat of formation for aspirin given the standard enthalpies of formation for CO_2, H_2O, and O_2 are -393.5 kJ/mol, -285.83 kJ/mol, and 0 kJ/mol, respectively.*

    This experiment does not consider that all of the conditions are standard state conditions; therefore, you are calculating ΔH_f **not** ΔH°_f.

3-10: Heat of Reaction: NaOH(aq) + HCl(aq)

Energy is either absorbed or released for all chemical reactions, and we call this energy the enthalpy of reaction (ΔH_{rxn}). If the enthalpy of reaction is positive, then we say that the energy was absorbed or that the reaction was *endothermic*. If the enthalpy of reaction is negative, then we say that energy was released or that the reaction was *exothermic*. Most chemical reactions are exothermic. In this problem, you will measure the amount of heat released when aqueous solutions of NaOH and HCl are mixed and react to form water and then you will calculate the heat of reaction.

$$NaOH(aq) + HCl(aq) = H_2O(l) + NaCl(aq)$$

1. Start *Virtual ChemLab,* select *Thermodynamics,* and then select *Heat of Reaction: NaOH(aq) + HCl(aq)* from the list of assignments. The lab will open in the Calorimetry laboratory.

2. Click the *Lab Book* to open it. In the thermometer window click **Save** to begin recording data. Allow 20-30 seconds to obtain a baseline temperature of the water. Pour the left beaker containing 100 mL of 1.000 M HCl into the calorimeter and then pour the right beaker containing 100 mL of 1.000 M NaOH into the calorimeter. Observe the change in temperature until it reaches a maximum and then record data for an additional 20-30 seconds. Click **Stop** in the temperature window. (You can click on the clock on the wall labeled *Accelerate* to accelerate the time in the laboratory.) A data link icon will appear in the lab book. Click the data link icon and record the temperature before adding the NaOH and the *highest* temperature after adding the NaOH in the data table. (Remember that the water will begin to cool down after reaching the equilibrium temperature.)

Data Table

	NaOH/HCl
initial temperature (°C)	
final temperature (°C)	

3. *Is the observed reaction endothermic or exothermic? What will be the sign of ΔH_{rxn}?* _____

_____

4. *Calculate the change in temperature, ΔT.* Record your results in the results table on the following page.

5. *Calculate the mass of the reaction mixture in the calorimeter.* (To do this, first determine the total volume of the solution based on the assumption that the volumes are additive and that the density of the solution is the same as that of pure water, 1.0 g/mL.) Record your results in the results table.

_____

6. *Calculate the total heat released in the reaction, assuming that the specific heat capacity of the solution is the same as that of pure water, 4.184 J/K·g.* Record the result in the results table. Remember: heat of reaction $= m \times C \times \Delta T$

_____

7. *Calculate the number of moles of NaOH used in the reaction by multiplying the volume of NaOH times the molarity (1.000 mol/L).* Record the results in the results table.

_____

8. *Calculate ΔH_{rxn}, in kJ/mol, of NaOH for the reaction and record the results in the results table.* Make sure the sign of ΔH_{rxn} is correct.

Results Table

Mass of Rxn Mixture	ΔT	Total Heat Released	mol NaOH	ΔH_{rxn}/mol

3-11: Heat of Reaction: MgO(s) + HCl(aq)

Energy is either absorbed or released for all chemical reactions, and we call this energy the enthalpy of reaction (ΔH_{rxn}). If the enthalpy of reaction is positive, then we say that the energy was absorbed or that the reaction was *endothermic*. If the enthalpy of reaction is negative, then we say that energy was released or that the reaction was *exothermic*. Most chemical reactions are exothermic. In this problem, you will measure the amount of heat released when solid MgO is reacted with aqueous HCl to form water and then you will calculate the heat of reaction.

$$MgO(s) + 2HCl(aq) = H_2O(l) + MgCl_2(aq)$$

1. Start *Virtual ChemLab*, select *Thermodynamics*, and then select *Heat of Reaction: MgO + HCl* from the list of assignments. The lab will open in the Calorimetry laboratory.

2. There will be a bottle of MgO near the balance. A weigh paper will be on the balance with approximately 2.81 g MgO on the paper. Record the mass of MgO in the data table.

3. The empty calorimeter will be on the lab bench and there will also be a beaker filled with 100 mL of 1.000 M HCl. Drag the beaker to the calorimeter until it snaps into place and pour the HCl into the calorimeter. Make certain the stirrer is *On* (you should be able to see the shaft rotating). Click the *Lab Book* to open it. In the thermometer window click *Save* to begin recording data. Allow 20-30 seconds to obtain a baseline temperature of the HCl solution. Drag the weigh paper containing the MgO sample over to the calorimeter and drop the sample in. Observe the change in temperature until it reaches a maximum and then record data for an additional 20-30 seconds. Click *Stop* in the temperature window. (You can click on the clock on the wall labeled *Accelerate* to accelerate the time in the laboratory.) A data link icon will appear in the lab book. Click the data link icon and record the temperature before adding the MgO and the *highest* temperature after adding the MgO in the data table. (Remember that the water will begin to cool down after reaching the equilibrium temperature.)

Data Table

	MgO/HCl
Mass MgO	
initial temperature (°C)	
final temperature (°C)	

4. *Is the observed reaction endothermic or exothermic? What will be the sign of ΔH_{rxn}?* _____

_____

5. *Calculate the change in temperature, ΔT. Record your results in the results table on the following page.*

6. *Calculate the mass of the reaction mixture in the calorimeter.* (To do this, assume that the density of the HCl solution originally in the calorimeter can be approximated with the density of water (1.0 g/mL). Record your results in the results table.

_____

7. *Calculate the total heat released in the reaction, in J, assuming that the specific heat capacity of the solution is the same as that of pure water, 4.184 J/K·g.* Record the result in the results table. Remember: heat of reaction = $m \times C \times \Delta T$

_____

8. *Calculate the number of moles of MgO used in the reaction.* The molar mass of MgO is 56.305 g/mole. Record the results in the results table.

_____

9. *Calculate ΔH_{rxn}, in kJ/mol, of MgO for the reaction and record the results in the results table.* Make sure the sign of ΔH_{rxn} is correct.

Results Table

Mass of Rxn Mixture	ΔT	Total Heat Released	mol MgO	ΔH_{rxn}/mol

3-12: Hess's Law

In this experiment, you will measure the amount of heat released in these three related exothermic reactions:

1. $NaOH\ (s) = Na^+\ (aq)\ +\ OH^-\ (aq)\ + \Delta H_1$
2. $NaOH\ (s)\ +\ H^+\ (aq)\ +\ Cl^-\ (aq) = H_2O\ +\ Na^+\ (aq)\ +\ Cl^-\ (aq)\ +\ \Delta H_2$
3. $Na^+\ (aq)\ +\ OH^-\ (aq) + H^+\ (aq)\ +\ Cl^-\ (aq) = H_2O\ +\ Na^+\ (aq)\ +\ Cl^-\ (aq)\ +\ \Delta H_3$

After determining the heats of reaction (ΔH_1, ΔH_2 and ΔH_3), you will then analyze your data and verify Hess's Law or $\Delta H_1 + \Delta H_3 = \Delta H_2$.

1. Start *Virtual ChemLab,* select *Thermodynamics,* and then select *Hess's Law* from the list of assignments. The lab will open in the Calorimetry laboratory.

Reaction 1

2. There will be a bottle of NaOH near the balance. A weigh paper will be on the balance with approximately 4 g NaOH on the paper. Record the mass of NaOH in the data table on the following page.

3. The calorimeter will be on the lab bench and filled with 200 mL water. Click the *Lab Book* to open it. Make certain the stirrer is *On* (you should be able to see the shaft rotating). In the thermometer window click *Save* to begin recording data. Allow 20-30 seconds to obtain a baseline temperature of the water.

4. Drag the weigh paper with the sample to the calorimeter until it snaps into place and then pour the sample into the calorimeter. Observe the change in temperature until it reaches a maximum and then record data for an additional 20-30 seconds. Click *Stop*. (You can click on the clock on the wall labeled *Accelerate* to accelerate the time in the laboratory.) A data link icon will appear in the lab book. Click the data link icon and record the initial and final water temperatures in the data table. If you need to repeat this part of the experiment, enter the Stockroom and select the preset experiment called *Heat of Solution-NaOH* on the clipboard.

Reaction 2

5. Click the red disposal bucket to clear the lab. Click on the Stockroom to enter. Click on the clipboard and select the preset experiment called *Heat of Reaction: HCl(aq) + NaOH(s)*. Return to the laboratory.

6. There will be a bottle of NaOH near the balance. A weigh paper will be on the balance with approximately 4 g NaOH on the paper. Record the mass of NaOH in the data table. The calorimeter will be on the lab bench filled with 100 mL water, and there will be a beaker containing 100 mL of 1.000 M HCl on the lab bench. In the thermometer window click *Save* to begin recording data. Allow 20-30 seconds to obtain a baseline temperature of the water.

7. Make sure the beaker of HCl is visible and drag it to the calorimeter and pour it into the calorimeter. The HCl and the water are at the same temperature so there should be no temperature change. Now drag the weigh paper with the NaOH to the calorimeter until it snaps into place and pour the sample into the calorimeter. It is important that the HCl be added first and the NaOH added second. Observe

the change in temperature until it reaches a maximum and then record data for an additional 20-30 seconds. Record the temperature before adding the HCl and the temperature after adding the NaOH in the data table.

Reaction 3

8. Click the red disposal bucket to clear the lab. Click on the Stockroom to enter. Click on the clipboard and select the preset experiment called *Heat of Reaction: HCl(aq) + NaOH(aq)*. Return to the laboratory.

9. In the thermometer window click *Save* to begin recording data. Allow 20-30 seconds to obtain a baseline temperature of the water. Pour the left beaker containing 100 mL of 1.0000M HCl into the empty calorimeter and then pour the right beaker containing 100 mL of 1.0000M NaOH into the calorimeter. Observe the change in temperature until it reaches a maximum and then record data for an additional 20-30 seconds. Record the initial and final temperatures in the data table.

Data Table

Parameter	Reaction 1	Reaction 2	Reaction 3
Mass NaOH			
initial temperature (°C)			
final temperature (°C)			

10. *Determine the change in temperature, ΔT, for each reaction.* Record your results in the results table on the following page.

11. *Calculate the mass of the reaction mixture in each reaction.* (To do this, first determine the total volume of the solution. Then calculate the mass of the solution, based on the assumption that the added solid does not change the volume and that the density of the solution is the same as that of pure water, 1.0 g/mL.) Remember to add the mass of the solid. Record your results in the result table.

12. *Calculate the total heat released in each reaction, in J, assuming that the specific heat capacity of the solution is the same as that of pure water, 4.184 J/K·g.* Record the result in the results table. Remember: heat of reaction = $m \times C \times \Delta T$

13. *Calculate the number of moles of NaOH used in reactions 1 and 2 where n = m/MW.* Record the results in the results table.

14. *Calculate the number of moles of NaOH used in reaction 3 by multiplying the volume of NaOH times the molarity (1.000 mol/L).* Record the results in the results table.

15. *Calculate the energy released, in kJ/mol, of NaOH for each reaction and record the results in the results table.*

    Reaction 1

    Reaction 2

    Reaction 3

Results Table

Rxn #	Mass of Rxn Mixture	ΔT	Total Heat Released	mol NaOH	Heat Released per mol NaOH
1					
2					
3					

16. *Show that the equations for reactions 1 and 3, which are given in the background section, add up to equal the equation for reaction 2. Include the energy released per mole of NaOH in each equation.*

17. *Calculate the percent difference between the heat given off in reaction 2 and the sum of the heats given off in reactions 1 and 3. Assume that the heat given off in reaction 2 is correct.*

$$\% \, Difference = \frac{\left| heat_2 - (heat_1 + heat_3) \right|}{heat_2} \times 100$$

% Difference =

18. *State in your own words what is meant by the additive nature of heats of reaction.* _____

    _____

    _____

    _____

3-13: The Balance Between Enthalpy and Entropy

For chemical reactions, we say that a reaction proceeds to the right when ΔG is negative and that the reaction proceeds to the left when ΔG is positive. At equilibrium ΔG is zero. The Gibbs-Helmholtz equation specifies that at constant temperature $\Delta G = \Delta H - T\Delta S$ or, in other words, that the sign and size of ΔG is governed by the balance between enthalpic (ΔH) and entropic (ΔS) considerations. In this assignment, you will dissolve several different salts in water, measure the resulting temperature changes, and then make some deductions about the thermodynamic driving forces behind the dissolving process.

1. Start *Virtual ChemLab,* select *Thermodynamics,* and then select *The Balance Between Enthalpy and Entropy* from the list of assignments. The lab will open in the Calorimetry laboratory.

2. There will be a bottle of sodium chloride (NaCl) on the lab bench. A weigh paper will be on the balance with approximately 2 g of NaCl on the paper.

3. The calorimeter will be on the lab bench and filled with 100 mL water. Click on the *Lab Book* to open it. Make certain the stirrer is *On* (you should be able to see the shaft rotating). In the thermometer window click *Save* to begin recording data. Allow 20-30 seconds to obtain a baseline temperature of the water.

4. Drag the weigh paper with the sample to the calorimeter until it snaps into place and then pour the sample in the calorimeter. Observe the change in temperature until it reaches a maximum (or minimum) and then record data for an additional 20-30 seconds. (You can click the clock on the wall labeled *Accelerate* to accelerate the time in the laboratory.) Click *Stop*. A data link icon will appear in the Lab Book. Click the data link icon and record the temperature before adding the NaCl and the *highest* or *lowest* temperature after adding the NaCl in the data table.

5. Click the red disposal bucket to clear the lab. Click on the *Stockroom* to enter. Click on the clipboard and select the preset experiment called *Heat of Solution-NaNO₃* and repeat the experiment with NaNO₃. Record the initial and final temperatures in the data table.

6. Click the red disposal bucket to clear the lab. Click on the *Stockroom* to enter. Click the clipboard and select the preset experiment called *Heat of Solution-NaAc* and repeat the experiment with NaCH₃COO (NaAc). Record the initial and final temperatures in the data table.

Data Table

Mixture	T_1	T_2	ΔT (T_2-T_1)
NaCl (s) + H$_2$O (l)			
NaNO$_3$ (s) + H$_2$O (l)			
NaCH$_3$COO + H$_2$O (l)			

Use your experimental data to answer the following questions:

7. *Calculate ΔT ($\Delta T = T_2$-T_1) for each mixture and record the results in the data table.*

8. An exothermic process releases heat (warms up), and an endothermic process absorbs heat (cools down).

    Which solutions are endothermic and which are exothermic? What is the sign of the change in enthalpy, ΔH, in each case?

    _____

9. *Which solution(s) had little or no change in temperature?* _____

10. When sodium chloride dissolves in water, the ions dissociate:

    $$NaCl\ (s)\ =\ Na^+\ (aq)\ +\ Cl^-\ (aq)$$

    Write ionic equations, similar to the one above, that describe how $NaNO_3$ and $NaCH_3COO$ each dissociate as they dissolve in water. Include heat as a reactant or product in each equation.

    _____

    _____

11. *What is the sign of the change in Gibbs free energy (ΔG) for each process?*

    _____

12. *Consider the Gibbs-Helmholtz equation, ΔG = ΔH - TΔS. For each dissolving process, substitute the signs of ΔG and ΔH into the equation and predict the sign for the entropy (ΔS). Does the sign for entropy change seem to make sense? Explain.*

    _____

    _____

    _____

    _____

13. *If the sign for ΔG is negative (spontaneous process) and the sign for ΔS is positive (more disorder) for both dissolving processes, how could one be endothermic (positive ΔH) and one be exothermic (negative ΔH)? Is there more to consider than just the dissolving process?*

    _____

    _____

    _____

    _____

    _____

    _____

    _____

    _____

4-1: Heat of Fusion of Water

The molar heat of fusion for a substance, ΔH_{fus}, is the heat required to transform one mole of the substance from the solid phase into the liquid phase. In this assignment, you will use a simple coffee cup calorimeter and a thermometer to measure the molar heat of fusion for water.

1. Start *Virtual ChemLab*, select *Colligative Properties,* and then select *Heat of Fusion of Water* from the list of assignments. The lab will open in the Calorimetry laboratory with a beaker of ice on the balance and a coffee cup calorimeter on the lab bench.

2. Click on the *Lab Book* to open it. Record the mass of the ice on the balance in the data table. If the mass is too small to read, click on the *Balance* area to zoom in. Note that the balance has already been tared for the mass of the empty beaker.

3. 100 mL of water is already in the coffee cup. Use the density of water at 25°C (0.998 g/mL) to determine the mass of water from the volume. Record the mass in the data table. Make certain the stirrer is **On** (you should be able to see the shaft rotating). In the thermometer window, click **Save** to begin recording data to the lab book. Allow 20-30 seconds to obtain a baseline temperature of the water.

4. Drag the beaker from the balance area until it snaps into place above the coffee cup and then pour the ice into the calorimeter. Click the thermometer and graph windows to bring them to the front and observe the change in temperature in the graph window until it reaches a minimum value and begins to warm again. Click **Stop** in the temperature window. (You can click on the clock on the wall labeled *Accelerate* to accelerate the time in the laboratory.) A data link icon will appear in the lab book. Click the data icon link and record the temperature before adding the ice and the *lowest* temperature after adding ice in the data table. (Remember that the water will begin to warm back up after all of the ice has melted.)

5. If you want to repeat the experiment, click on the red disposal bucket to clear the lab, click on the Stockroom, click on the clipboard, and select Preset Experiment #3, Heat of Fusion of Water.

Data Table

volume of water in calorimeter (mL)	
mass of water in calorimeter (g)	
mass of ice (g)	
initial temperature (°C)	
final temperature (°C)	

6. Calculate ΔT for the water using $\Delta T = |T_f - T_i|$. _____

7. Calculate the heat (q) transferred, in kJ, from the water to the ice using $q = mHCH\Delta T$ where the heat capacity (C) for water is 4.18 J/K·g and the mass, m, is for the water in the calorimeter.

_____

8. *Convert the mass of ice to moles.* _____

9. *Calculate ΔH_{fus} of water, in kJ/mol, by dividing the heat transferred from the water by the moles of*

   ice melted. _____

10. *Compare your experimental value of ΔH_{fus} of ice with the accepted value of 6.01 kJ/mol and calculate the % error using the formula:*

$$\% \ Error = \frac{|your \ answer - accepted \ answer|}{accepted \ answer} \times 100$$

   % Error =

11. *What are some possible sources of error in this laboratory procedure? ?* _____

   _____

   _____

   _____

4-2: Heat of Vaporization of Water

A graph of the vapor pressure of a liquid as a function of temperature has a characteristic shape that can be represented fairly accurately with the *Clausius-Clapeyron* equation. In this assignment you will measure the vapor pressure of water at several temperatures and then use this data to estimate the heat of vaporization of water using the *Clausius-Clapeyron* equation.

1. Start *Virtual ChemLab*, select *Colligative Properties*, and then select *Heat of Vaporization of Water* from the list of assignments. The lab will open in the Gases laboratory.

2. The balloon is filled with 0.10 moles of water vapor at a pressure of 140 kPa and a temperature of 400 K. Pull down the lever on the Temperature LCD controller until the temperature stops decreasing. This temperature represents the equilibrium temperature where water as a gas exists in equilibrium with water as a liquid. The pressure at this temperature is the vapor pressure. Record the vapor pressure (in kPa) and the temperature (in K) in the data table.

3. Change the pressure from 140 kPa to 120 kPa using the lever on the Pressure LCD controller or by clicking on the tens place and typing "2." Pull down the lever on the Temperature controller until the temperature stops decreasing. Record the vapor pressure (in kPa) and the temperature (in K) in the data table. Continue this process with pressures of 100 kPa, 80 kPa, 60 kPa, and 40 kPa. Do not forget to decrease the temperature until the temperature stops decreasing after each pressure change. Record all vapor pressure and temperatures in the data table.

Data Table

Vapor Pressure (kPa)	Temperature (K)

4. The *Clausius-Clapeyron* equation has the form $\ln\dfrac{P_2}{P_1} = -\dfrac{\Delta H_{vap}}{R}\left(\dfrac{1}{T_2} - \dfrac{1}{T_1}\right)$. If you plot ln P (natural log of the vapor pressure) as a function of $1/T$, the data should form a straight line. Using a spreadsheet program and the data you collected, plot $1/T$ (reciprocal temperature) on the x-axis and ln P (natural log of the vapor pressure) on the y-axis.

Describe the curve. _____

5. Graphically or using the spreadsheet program, determine the best linear fit for your curve and find the slope of the line. From the *Clausius-Clapeyron* equation, ΔH_{vap} = -slope × R where R = 8.314 J·K^{-1}·mol^{-1}.

Based on your slope, calculate the heat of vaporization of water in kJ/mol.

_____

6. The accepted value for the heat of vaporization of water is 40.7 kJ/mol.

    Calculate the percent error using the formula:

    $$\% \, Error = \frac{|your \, answer - accepted \, answer|}{accepted \, answer} \times 100$$

    % Error =

4-3: The Boiling Point of Water at High Altitude

The relationship between the equilibrium vapor pressure of a liquid or solid and temperature is given by the *Clausius-Clapeyron* equation. In this assignment you will measure the vapor pressure of water at a given temperature and use this data and the *Clausius-Clapeyron* equation to calculate the boiling point of water on the top of Mt. Denali in Alaska.

1. Start *Virtual ChemLab,* select *Colligative Properties,* and then select *The Boiling Point of Water at High Altitude* from the list of assignments. The lab will open in the Gases laboratory.

2. The balloon is filled with 0.40 moles of water vapor at a pressure of 1500 Torr and a temperature of 400 K. Pull down the lever on the Temperature LCD controller until the temperature stops decreasing. This temperature represents the equilibrium temperature where water as a gas exists in equilibrium with water as a liquid. The pressure at this temperature is the vapor pressure. Record the vapor pressure (in Torr) and the temperature (in K) in the data table.

Data Table

Vapor Pressure (Torr)	Temperature (K)

3. The *Clausius-Clapeyron* equation may be written in several forms. For this assignment, the most useful form can be written as

$$\ln\frac{P_2}{P_1} = -\frac{\Delta H_{vap}}{R}\left(\frac{1}{T_2} - \frac{1}{T_1}\right)$$

If P_1 and T_1 are the experimental vapor pressure and temperature that you measured and the pressure at the top of Mt. Denali, P_2, is 340 Torr, the boiling point of water, T_2, at the top of the mountain can be calculated by solving the *Clausius-Clapeyron* equation for T_2.

Calculate the boiling point of water at the top of Mt. Denali. The value of R is 8.314 J·K^{-1}·mol^{-1} and ΔH_{vap} for water is 40.67 kJ/mol

4-4 Boiling Point Elevation

If you dissolve a substance such as ordinary table salt (NaCl) in water, the boiling point of the water will increase relative to the boiling point of the pure water. In this assignment, you will dissolve a sample of NaCl in water and then measure the boiling point elevation for the solution.

1. Start *Virtual ChemLab, select Colligative Properties,* and then select *Boiling Point Elevation* from the list of assignments. The lab will open in the Calorimetry laboratory with a calorimeter on the lab bench and a sample of sodium chloride (NaCl) on the balance.

2. Record the mass of the sodium chloride in the data table. If it is too small to read, click on the *Balance* area to zoom in, record the reading, and then return to the laboratory.

3. 100 mL of water is already in the calorimeter. Use the density of water at 25°C (0.998 g/mL) to determine the mass from the volume and record it in the data table. Make certain the stirrer is *On* (you should be able to see the shaft rotating). Click on the green heater light on the control panel to turn on the heater and begin heating the water. Click the clock on the wall labeled *Accelerate* to accelerate the laboratory time if necessary.

4. Observe the temperature until the first appearance of steam comes from the calorimeter. Immediately click the red light on the heater to turn it off and then record the temperature as the boiling point of pure water in the data table. Letting the water boil will decrease the mass of the water present in the calorimeter. Note that the boiling point may be different than 100°C if the atmospheric pressure is *not* 760 Torr. The current atmospheric pressure for the day can be checked by selecting *Pressure* on the LED meter on the wall.

5. Drag the weigh paper to the calorimeter and add the NaCl. Wait 30 seconds for the salt to dissolve and then turn on the heater. When steam first appears observe and record the temperature in the data table.

6. If you want to repeat the experiment, click on the red disposal bucket to clear the lab, click on the Stockroom, click on the clipboard, and select Preset Experiment #2, Boiling Point Elevation – NaCl.

Data Table

mass NaCl	
mass water	
boiling temp of pure water	
boiling temp of solution	

7. The boiling point elevation can be predicted using the equation $\Delta T = K_b \times m \times i$, where ΔT is the change in boiling point, i is the number of ions in the solution per mole of dissolved NaCl ($i = 2$), m is the molality of the solution, and K_b is the molal boiling point constant for water which is 0.51°C/m.

Calculate the predicted change in boiling point, in EC for your solution.

8. The change in boiling point must be added to the boiling point of pure water in your experiment in order to compare the predicted boiling point with the actual boiling point.

What is the calculated boiling point of the solution? Compare this with the actual boiling point.

_____

_____

4-5: Freezing Point Depression

If you dissolve a substance such as ordinary table salt (NaCl) in water, the freezing point of the water will decrease relative to the freezing point of the pure water. This property is used to melt the snow or ice on roads during the winter or to make homemade ice cream. In this assignment, you will dissolve a sample of NaCl in water, add some ice, and then measure the freezing point depression.

1. Start *Virtual ChemLab,* select *Colligative Properties,* and then select *Freezing Point Depression* from the list of assignments. The lab will open in the Calorimetry laboratory with a beaker containing 45.00 g of ice and a coffee cup calorimeter on the lab bench. A sample of sodium chloride (NaCl) will also be on the balance.

2. Click on the *Lab Book* to open it. Record the mass of the sodium chloride in the data table. If it is too small to read, click on the *Balance* area to zoom in, record the mass, and then return to the laboratory.

3. 100 mL of water is already in the calorimeter. Use the density of water at 25°C (0.998 g/mL) to determine the mass from the volume and record it in the data table. Make certain the stirrer is *On* (you should be able to see the shaft rotating). In the thermometer window, click *Save* to begin recording data in the lab book. Allow 20-30 seconds to obtain a baseline temperature of the water. Click the clock on the wall labeled *Accelerate* to accelerate the laboratory time if necessary.

4. Drag the beaker of ice until it snaps into place above the calorimeter and then pour the ice into the calorimeter. Click the thermometer and graph windows to bring them to the front and observe the change in temperature in the graph window until it reaches zero. Drag the weigh paper from the balance to the calorimeter and then pour it into the calorimeter. Observe the change in temperature until it reaches a stable minimum and click *Stop* in the temperature window. A data link icon will appear in the lab book. Click the data link icon and record the *lowest* temperature after adding the salt in the data table. (Remember that the water may have begun to warm back up.)

5. If you want to repeat the experiment, click on the red disposal bucket to clear the lab, click on the Stockroom, click on the clipboard, and select Preset Experiment #1, Freezing Point Depression – NaCl.

Data Table

mass NaCl	
mass water	
mass ice	
mass water + ice	
minimum temperature	

6. The freezing point depression can be predicted using the equation $\Delta T = K_f \times m \times i$, where ΔT is the change in freezing point, i is the number of ions in the solution per mole of dissolved NaCl ($i = 2$), m is the molality of the solution, and K_f is the molal freezing point constant for water which is 1.86°C/m.

Calculate the predicted change in freezing point for your solution.

7. The change in freezing point must be subtracted from the freezing point of pure water, which is 0.0 °C, in order to compare the predicted freezing point with the actual freezing point.

What is the calculated freezing point of the solution? Compare this to the actual freezing point.

4-6: Molar Mass Determination by Boiling Point Elevation

If you dissolve a substance such as ordinary table salt (NaCl) in water, the boiling point of the water will increase relative to the boiling point of the pure water. You can use this property to calculate the molar mass of an unknown. In this assignment, you will dissolve a sample of NaCl in water, measure the boiling point elevation for the solution, and then calculate the molar mass for NaCl as if it were an unknown.

1. Start *Virtual ChemLab,* select *Colligative Properties,* and then select *Molar Mass Determination by Boiling Point Elevation* from the list of assignments. The lab will open in the Calorimetry laboratory with a calorimeter on the lab bench and a sample of sodium chloride (NaCl) on the balance.

2. Record the mass of the sodium chloride in the data table. If it is too small to read, click on the *Balance* area to zoom in, record the reading, and then return to the laboratory.

3. 100 mL of water is already in the calorimeter. Use the density of water at 25°C (0.998 g/mL) to determine the mass from the volume and record it in the data table. Make certain the stirrer is **On** (you should be able to see the shaft rotating). Click on the green heater light on the control panel to turn on the heater and begin heating the water. Click the clock on the wall labeled *Accelerate* to accelerate the laboratory time if necessary.

4. Observe the temperature until the first appearance of steam comes from the calorimeter. Immediately click the red light on the heater to turn it off and then record the temperature as the boiling point of pure water in the data table. Letting the water boil will decrease the mass of the water present in the calorimeter. Note that the boiling point may be different than 100°C if the atmospheric pressure is *not* 760 Torr. The current atmospheric pressure for the day can be checked by selecting *Pressure* on the LED meter on the wall.

5. Drag the weigh paper to the calorimeter and add the NaCl. Wait 30 seconds for the salt to dissolve and then turn on the heater. When steam first appears, observe and record the temperature in the data table.

6. If you want to repeat the experiment, click on the red disposal bucket to clear the lab, click on the Stockroom, click on the clipboard, and select Preset Experiment #2, Boiling Point Elevation – NaCl.

Data Table

mass NaCl	
mass water	
boiling temp of pure water	
boiling temp of solution	

7. *Calculate the boiling point elevation, ΔT, caused by adding NaCl to the water.*

_____

8. The boiling point elevation can be calculated using the equation $\Delta T = K_b \times m \times i$, where ΔT is the change in boiling point, i is the number of ions in the solution per mole of dissolved NaCl ($i = 2$), m is the molality of the solution, and K_b is the molal boiling point constant for water which is 0.51°C/m.

Using this equation and the data recorded in the data table, calculate the molar mass for NaCl and compare it with the actual value.

4-7: Molar Mass Determination by Freezing Point Depression

If you dissolve a substance such as ordinary table salt (NaCl) in water, the freezing point of the water will decrease relative to the freezing point of the pure water. You can use this property to calculate the molar mass of an unknown. In this assignment, you will dissolve a sample of NaCl in water, measure the freezing point depression for the solution, and then calculate the molar mass for NaCl as if it were an unknown.

1. Start *Virtual ChemLab, select Colligative Properties,* and then select *Molar Mass Determination by Freezing Point Depression* from the list of assignments. The lab will open in the Calorimetry laboratory with a beaker containing 45.00 g of ice and a coffee cup calorimeter on the lab bench. A sample of sodium chloride (NaCl) will also be on the balance.

2. Click on the *Lab Book* to open it. Record the mass of the sodium chloride in the data table. If it is too small to read, click on the *Balance* area to zoom in, record the mass, and then return to the laboratory.

3. 100 mL of water is already in the calorimeter. Use the density of water at 25°C (0.998 g/mL) to determine the mass from the volume and record it in the data table. Make certain the stirrer is *On* (you should be able to see the shaft rotating). IN the thermometer window, click *Save* to begin recording data in the lab book. Allow 20-30 seconds to obtain a baseline temperature of the water. Click the clock on the wall labeled *Accelerate* to accelerate the laboratory time if necessary.

4. Drag the beaker of ice until it snaps into place above the calorimeter and then pour the ice into the calorimeter. Click the thermometer and graph windows to bring them to the front and observe the change in temperature in the graph window until it reaches zero. Drag the weigh paper from the balance to the calorimeter and then pour it into the calorimeter. Observe the change in temperature until it reaches a stable minimum and click *Stop* in the temperature window. A data link icon will appear in the lab book. Click the data link icon and record the *lowest* temperature after adding the salt in the data table. (Remember that the water may have begun to warm back up.)

5. If you want to repeat the experiment, click on the red disposal bucket to clear the lab, click on the Stockroom, click on the clipboard, and select Preset Experiment #1, Freezing Point Depression – NaCl.

Data Table

mass NaCl	
mass water	
mass ice	
mass water + ice	
minimum temperature	

6. *What is the freezing point depression caused by adding NaCl to the water?* _____

7. The freezing point depression can be calculated using the equation $\Delta T = K_f \times m \times i$, where ΔT is the change in freezing point, i is the number of ions in the solution per mole of dissolved NaCl ($i = 2$), m is the molality of the solution, and K_f is the molal freezing point constant for water which is 1.86°C/m.

Using this equation and the data recorded in the data table, calculate the molar mass for NaCl and compare it with the actual value.

_____

4-8: Changes in the Boiling Point

If you dissolve a substance such as ordinary table salt (NaCl) in water, the boiling point of the water will increase relative to the boiling point of the pure water. In this assignment, you will dissolve a sample of NaCl in water and then observe what happens when you continue to boil the water until the water boils away.

1. Start *Virtual ChemLab,* select *Colligative Properties,* and then select *Changes in the Boiling Point* from the list of assignments. The lab will open in the Calorimetry laboratory with a calorimeter on the lab bench and a sample of sodium chloride (NaCl) on the balance.

2. 100 mL of water is already in the calorimeter. Make certain the stirrer is ***On*** (you should be able to see the shaft rotating). Drag the weigh paper to the calorimeter and add the NaCl. Wait 30 seconds for the salt to dissolve and then turn on the heater by clicking on the green heater light on the control panel. Click the clock on the wall labeled *Accelerate* to accelerate the laboratory time if necessary.

3. When steam first appears, observe the temperature. Allow the solution to continue boiling and observe the temperature until all of the water boils away and the heater burns out. (The water level is shown on the right side of the calorimeter control panel.)

    What observations did you make about the temperature as the solution continued to boil?

    _____

4. *Explain the observations described in question # 3. Why does this occur?*

    _____

    _____

    _____

5-1: Boyle's Law: Pressure and Volume

Robert Boyle, a philosopher and theologian, studied the properties of gases in the 17th century. He noticed that gases behave similarly to springs; when compressed or expanded, they tend to 'spring' back to their original volume. He published his findings in 1662 in a monograph entitled *The Spring of the Air and Its Effects*. You will make observations similar to those of Robert Boyle and learn about the relationship between the pressure and volume of an ideal gas.

1. Start *Virtual ChemLab,* select *Gas Properties,* and then select *Boyle's Law: Pressure and Volume* from the list of assignments. The lab will open in the Gases laboratory.

2. Note that the balloon in the chamber is filled with 0.300 moles of an ideal gas (MW = 4 g/mol) at a temperature of 298 K, a pressure of 1.00 atm, and a volume of 7.336 L. To the left of the Pressure LCD controller is a lever that will decrease and increase the pressure as it is moved up or down; the digit changes depending on how far the lever is moved. Digits may also be clicked directly to type in the desired number. You may want to practice adjusting the lever so that you can decrease and increase the pressure accurately. Make sure the moles, temperature, and pressure are returned to their original values before proceeding.

3. Click on the *Lab Book* to open it. Back in the laboratory, click on the *Save* button to start recording *P, V, T,* and *n* data to the lab book. Increase the pressure from 1 atm to 10 atm one atmosphere at a time. Click *Stop* to stop recording data, and a data link icon will appear in the lab book. To help keep track of your data links, enter 'Ideal Gas 1' next to the link.

4. *Zoom Out* by clicking the green arrow next to the *Save* button. Click *Return Tank* on the gas cylinder. On the table underneath the experimental chamber is a switch to choose Real gases or Ideal gases. Click on the *Ideal Gases* and choose the cylinder labeled *Ideal 8* (Ideal 8 MW = 222 g/mol). Click on the balloon chamber to *Zoom In* and set the temperature, pressure, and moles to 298 K, 1.00 atm, and 0.300 moles, respectively. Repeat the experiment with this gas labeling the data link as 'Ideal Gas 8.'

5. *Zoom Out* by clicking on the green arrow next to the *Save* button. Click on the *Stockroom* and then on the *Clipboard* and select *Balloon Experiment N2*. Again, set the temperature, pressure, and moles to 298 K, 1.00 atm, and 0.300 moles, respectively. You may have to click on the *Units* button to change some of the variables to the correct units. Repeat the experiment with this gas labeling the data link 'Real Gas N2.'

6. Select the lab book and click on the data link iocn for Ideal Gas 1. In the *Data Viewer* window, select all the data by clicking on the *Copy Data* button and copy the data using CTRL-C for Windows or CMD-C for Macintosh. Paste the data into a spreadsheet program and create a graph with volume on the *x*-axis and pressure on the *y*-axis. Also create a graph for your data from Ideal Gas 8 and Real Gas N2.

7. *Based on your data, what relationship exists between the pressure and the volume of a gas (assuming*

    a constant temperature)? _____

_____

8. *Look up a statement of Boyle's Law in your textbook. Do your results further prove this?* _____

9. Complete the tables from the data saved in your lab book. Use only a sampling of the data for pressures at 1, 3, 6, and 9 atm.

Ideal Gas 1 MW = 4 g/mol

Volume (L)	Pressure (atm)	PV Product ($P \times V$)

Ideal Gas 8 MW = 222 g/mol

Volume (L)	Pressure (atm)	PV Product ($P \times V$)

Real Gas N2

Volume (L)	Pressure (atm)	PV Product ($P \times V$)

10. *What conclusions can you make about the PV product with Ideal Gas 1, MW = 4 g/mol?*

_____

How is the PV product affected using an ideal gas with a different molecular weight (Ideal Gas 8)?

_____

11. *How are your results affected using a Real Gas (N2)?* _____

_____

12. You may want to repeat the experiment several times using different size pressure changes.

5-2: Charles' Law: Temperature and Volume

Charles' Law was discovered by Joseph Louis Gay-Lussac in 1802; it was based on unpublished work done by Jacques Charles in about 1787. Charles had found that a number of gases expand to the same extent over the same 80 degree temperature interval. You will be observing the relationship between the temperature and volume of a gas similar to that studied by Charles.

1. Start *Virtual ChemLab, select Gas Properties,* and then select *Charles's Law: Temperature and Volume* from the list of assignments. The lab will open in the Gases laboratory.

2. Note that the balloon in the chamber is filled with 0.050 moles of an ideal gas (MW = 4 g/mol) at a temperature of 100°C, a pressure of 1.00 atm, and a volume of 1.531 L. To the left of the Temperature LCD controller is a lever that will decrease and increase the temperature as it is moved up or down; the digit changes depending on how far the lever is moved. Digits may also be clicked directly to type in the desired number, or they can be rounded by clicking on the *R* button. You may want to practice adjusting the lever so that you can decrease and increase the temperature accurately. Make sure the moles, temperature, and pressure are returned to their original values before proceeding.

3. Click on the *Lab Book* to open it. Back in the laboratory, click on the *Save* button to start recording *P, V, T,* and *n* data to the lab book. Increase the temperature from 100°C to 1000°C 100 degrees at a time. Click *Stop* to stop recording data, and a data link icon will appear in the lab book. To help keep track of your data links, enter 'Ideal Gas 1' next to the link.

4. *Zoom Out* by clicking on the green arrow next to the *Save* button. Click on the *Stockroom* and then on the *Clipboard* and select *Balloon Experiment N2*. Again, set the temperature, pressure, and moles to 100°C, 1.00 atm, and 0.050 moles, respectively. You may have to click on the *Units* button to change some of the variables to the correct units. Repeat the experiment with this gas labeling the data link 'Real Gas N2.'

5. Select the lab book and click on the data link icon for Ideal Gas 1. In the *Data Viewer* window, select all the data by clicking on the *Copy Data.* Paste the data into a spreadsheet program and create a graph with temperature on the *x*-axis and volume on the *y*-axis. Also create a graph for the data labeled Real Gas N2.

6. *Based on your data, what relationship exists between the temperature and the volume of a gas (assuming a constant pressure)?* _____

    _____

7. *Look up a statement of Charles' Law in your textbook. Do your results further prove this?* _____

8. Using the spreadsheet program, fit the ideal gas data to a line or printout the graph and use a ruler to draw the best line through the data. The lowest possible temperature is reached when an ideal gas has zero volume. This temperature is the *x*-intercept for the plotted line.

    What is this temperature? _____

9. *Now do the same analysis with the real gas data (N2). What temperature did you find?* _____

10. *Under these conditions, does N₂ behave like an ideal gas?* _____

5-3: Avogadro's Law: Moles and Volume

In 1808, Joseph Gay-Lussac observed the *law of combining values,* which states that the volumes of gases that react with one another react in the ratio of small whole numbers. Three years later, Amedeo Avogadro built upon this observation by proposing what is now known as *Avogadro's hypothesis*: Equal volumes of gases at the same temperature and pressure contain equal numbers of molecules. Avogadro's Law, which states the relationship between moles and volume, followed from his hypothesis. You will be observing the same principle that Avogadro stated two hundred years ago.

1. Start *Virtual ChemLab,* select *Gas Properties,* and then select *Avogadro's Law: Moles and Volume* from the list of assignments. The lab will open in the Gases laboratory. You will see LCD controllers on the left for volume, pressure, temperature and number of moles. You may change the units for volume, pressure, and temperature by clicking on the *Units* buttons. The balloon has been filled with 0.100 moles of an ideal gas and the pressure is 2.00 atm. Record the number of moles and the volume (in L) in the data table below.

2. Click on the tenths digit on the Moles LCD controller and change the moles of gas in the balloon from 0.1 to 0.2 mole. Record the number of moles and volume in the data table. Repeat for 0.3, 0.4, and 0.5 moles.

DataTable

n (moles)	V (L)

3. *Based on your observations, what can you state about the relationship between moles and volume*

   of a gas? _____

   _____

4. *Write a mathematical equation using a proportionality constant (k) with units of L/mol that expresses what you have learned about Avogadro's Law. Determine the value of k.*

5-4: Derivation of the Ideal Gas Law

An ideal gas is a hypothetical gas whose pressure, volume, and temperature follow the relationship $PV = nRT$. Ideal gases do not actually exist, although all real gases can behave like an ideal gas at certain temperatures and pressures. All gases can be described to some extent using the Ideal Gas Law, and it is important in our understanding of how all gases behave. In this assignment, you will derive the Ideal Gas Law from experimental observations.

The state of any gas can be described using the four variables: pressure (P), volume (V), temperature (T), and the number of moles of gas (n). Each experiment in *Virtual ChemLab: Gases* allows three of these variables (the independent variables) to be manipulated or changed and shows the effect on the remaining variable (the dependent variable).

1. Start *Virtual ChemLab,* select *Gas Properties,* and then select *Ideal Gas Law* from the list of assignments. The lab will open in the Gases laboratory.

2. Use the balloon experiment already setup in the laboratory to describe the relationship between pressure (P) and volume (V). Increase and decrease the pressure using the lever on the Pressure LCD controller to determine the effect on volume.

   What can you conclude about the effect of pressure on volume? Write a mathematical relationship using the proportionality symbol ($\propto$).

   _____

3. Use this same experiment to describe the relationship between temperature (T) and volume by increasing and decreasing the temperature.

   What can you conclude about the effect of temperature on volume? Write a mathematical relationship using the proportionality symbol ($\propto$).

   _____

4. Use this same experiment to describe the relationship between moles of gas and volume by increasing and decreasing the number of moles (n).

   What can you conclude about the effect of moles on volume? Write a mathematical relationship using the proportionality symbol ($\propto$).

   _____

5. Since volume is inversely proportional to pressure and directly proportional to temperature and moles, we can combine these three relationships into a single proportionality by showing how V is proportional to $1/P$, T, and n.

Write one combined proportion to show the relationship of volume to pressure, temperature and moles.

_____

6. This proportional relationship can be converted into a mathematical equation by inserting a proportionality constant (R) into the numerator on the right side.

   Write this mathematical equation and rearrange with P on the left side with V.

_____

7. This equation is known as the Ideal Gas Law.

   Using data for volume, temperature, pressure and moles from one of the gas experiments, calculate the value for R with units of $L \cdot atm \cdot K^{-1} \cdot mol^{-1}$. (Show all work and round to three significant digits.)

_____

8. *Using the conversion between atmospheres and mm Hg (1 atm = 760 mm Hg), calculate the value for R with units of $L \cdot mm\ Hg \cdot K^{-1} \cdot mol^{-1}$. (Show all work and round to three significant digits.)*

_____

9. *Using the conversion between atmospheres and kPa (1 atm = 101.3 kPa), calculate the value for R with units of $L \cdot kPa \cdot K^{-1} \cdot mol^{-1}$. (Show all work and round to three significant digits.)*

_____

5-5: Dalton's Law of Partial Pressures

Dalton's Law of Partial Pressures, named after its discoverer John Dalton, describes the behavior of gas mixtures. It states that the total pressure of the gas, P_{tot}, is the sum of the partial pressures of each gas, or the sum of the pressures that each gas would exert if it were alone in the container. In this assignment you will become more familiar with Dalton's law.

1. Start *Virtual ChemLab,* select *Gas Properties,* and then select *Dalton's Law of Partial Pressures* from the list of assignments. The lab will open in the Gases laboratory. You will see a gas experiment with eight gas cylinders on the right. Make certain that the switch on the lower right of the lab bench is set to Ideal Gases. Note that the Ideal Gases each have a different molecular weight.

2. Select one of the Ideal Gas cylinders by clicking on the white label. Click the red arrow on the brass cylinder valve until the meter reads 400. Add this Ideal Gas to the balloon by clicking the green **Open Valve** switch once to add gas and again to stop. Add an amount of your choice but do not fill the balloon too full since you will be adding two additional gases. Click **Return Tank**.

3. Repeat step # 2 for two additional Ideal Gases of your choice. Make certain that you do not explode the balloon. If you do, click the **Reset** button located on the upper right of the gas chamber and repeat the experiment.

4. Click on the experimental apparatus to *Zoom In*. There are four LCD controllers on the left for volume, pressure, temperature, and number of moles. You can change the units for volume, pressure, and temperature by clicking on the **Units** button on each controller. Under pressure and number of moles are numbers 1-8 that correspond to Ideal Gases 1-8. The three gases that you selected will be highlighted. Clicking on each highlighted number will display the pressure or moles for that gas alone. Find the Ideal Gas number, the number of moles, and the partial pressure for each of your three Ideal Gases. Record this in the data table.

Data Table

Ideal Gas Number	Moles (n)	Partial Pressure (P_i)

5. *Using the information from the data table, determine the total pressure in the balloon.*

_____

6. Click **Total** on the Pressure controller. Compare your answer from # 5 to the total pressure on the meter.

    Write both pressures below and write a mathematical equation to represent what you have learned about Dalton's Law.

7. Another way of expressing Dalton's Law of Partial Pressures is with the expression $P_i = x_i P_{total}$ where P_i is the partial pressure of gas i, x_i is the mole fraction of that gas in the gas mixture, and P_{total} is the total pressure.

    Verify that this relationship holds using the data you have collected and record you results in the data table below.

    ### Data Table

Ideal Gas Number	x_i	P_i (Calculated)	P_i (Measured)

5-6: Ideal vs. Real Gases

At room temperature and normal atmospheric pressures, real gases behave similarly to ideal gases, but real gases can deviate significantly from ideal gas behavior at extreme conditions such as high temperatures and high pressures. It is not always easy to find an effective means to show the deviation from ideal gas behavior for real gases. An effective but simple method to see these deviations is to calculate the value of R, the ideal gas constant, for real gases *assuming* they follow the ideal gas law and then compare the value of R with the actual value of 0.08205 L·atm·K^{-1}·mol^{-1}. In this assignment, you will measure P, V, T, and n for various real gases under different conditions and then you will calculate the value of R and compare it with its actual value.

1. Start *Virtual ChemLab,* select *Gas Properties,* and then select *Ideal vs. Real Gases* from the list of assignments. The lab will open in the Gases laboratory with the balloon filled with 0.100 moles of Ideal Gas 1.

2. Click on the ***Units*** buttons to change the units to L or mL for volume, atm for pressure, and K for temperature. On the left of the gas chamber are LCD controllers for the volume, pressure, temperature, and moles. On the left of the Pressure, Temperature, and Moles LCD controllers is a lever that will decrease and increase the pressure, temperature, or moles as the lever is moved up or down; the digit changes depending on how far the lever is moved. Digits may also be clicked directly to type in the desired number. Clicking to the left of the farthest left digit will add the next place; for example, if you have 1.7 atm you can click left of the 1 and enter 2 to make it 21.7 atm or click left of the 2 and enter 5 to make it 521.7 atm. The small ***R*** button in the upper left corner rounds the number. Clicking several times will round from ones to tens to hundreds.

   The green arrow to the left of the ***Save*** button will *Zoom Out*. Clicking ***Return Tank*** on the gas cylinder will return the tank to the rack and allow you to select a different gas. Clicking the gas chamber will *Zoom In* to allow you to change parameters. Be careful not to make the balloon so large that it bursts. If it does, click the red ***Reset*** button in the top right and then reset your units and values for each parameter. Remember that volume must be in L. If mL appears, you must convert to L in your calculations.

3. *Complete the data table for the following gases and conditions (all with 0.100 mole):*

   a. Ideal gas at low $T = 10$ K, high $T = 1000$ K, low $P = 1$ atm, high $P = 15$ atm
   b. Methane gas (CH_4) at low $T = 160$ K, high $T = 400$ K, low $P = 1$ atm, high $P = 15$ atm
   c. Carbon dioxide gas (CO_2) at low $T = 250$ K, high $T = 1000$ K, low $P = 1$ atm, high $P = 15$ atm

Data Table

Gas	V (L)	P (atm)	T (K)	n (mol)
Ideal, low T, low P				
Ideal, low T, high P				
Ideal, high T, low P				
Ideal, high T, high P				
CH_4, low T, low P				
CH_4, low T, high P				
CH_4, high T, low P				
CH_4, high T, high P				

CO_2, low T, low P				
CO_2, low T, high P				
CO_2, high T, low P				
CO_2, high T, high P				

4. If $PV = nRT$ then $R = PV/nT$.

   Complete the results table for each experiment above. Use four significant digits.

Results Table

Gas	Calculated R ($L \cdot atm \cdot K^{-1} \cdot mol^{-1}$)
Ideal, low T, low P	
Ideal, low T, high P	
Ideal, high T, low P	
Ideal, high T, high P	
CH_4, low T, low P	
CH_4, low T, high P	
CH_4, high T, low P	
CH_4, high T, high P	
CO_2, low T, low P	
CO_2, low T, high P	
CO_2, high T, low P	
CO_2, high T, high P	

5. *Which gases and conditions show significant deviation from the actual value of R? Explain.*

   _____

   _____

   _____

   _____

5-7: The Effect of Mass on Pressure

An understanding of pressure is an integral part of our understanding of the behavior of gases. Pressure is defined as the force per unit area exerted by a gas or other medium. The pressure of a gas is affected by many variables, such as temperature, external pressure, volume, moles of a gas, and other factors. This assignment will help you become more familiar with pressure and the effect of adding mass to a frictionless, massless piston.

1. Start *Virtual ChemLab,* select *Gas Properties,* and then select *The Effect of Mass on Pressure* from the list of assignments. The lab will open in the Gases laboratory.

2. This experiment consists of a cylinder with a frictionless-massless piston. When the experiment starts, the chamber is filled with the selected gas. Clicking on the **Piston** button moves the piston onto the cylinder and traps the gas in the cylinder. The moles of gas trapped in the cylinder and the volume of gas are measured using the Moles/Volume LCD controller. Pressure can be exerted on the gas in the cylinder by adjusting the external pressure in the chamber and by adding mass to the top of the piston (or $P_{int} = P_{mass} + P_{ext}$ where P_{int} is the internal pressure or the pressure of the gas in the cylinder, P_{mass} is the pressure being exerted on the gas by adding weights to the piston, and P_{ext} is the pressure being exerted on the piston by the gas in the chamber). If there is no mass on the piston, then $P_{int} = P_{ext}$.

3. Click the green **Piston** button to move the piston onto the cylinder. Record the mass (force, in tons) and the internal pressure (in psi) in the data table.

4. Click on the tenths place for mass and add 0.5 tons of mass to the piston. Record the mass and internal pressure in the Data Table. Repeat this for 2.5 tons (the weight of a small car).

Data Table

Mass (tons)	External Pressure (psi)	Calculated Internal Pressure (psi)	Measured Internal Pressure (psi)

5. You must now calculate the pressure being exerted by the 0.5 tons or, P_{mass}. First, convert tons to psi (pounds per square inch).

    How many pounds is 0.5 tons? _____

    The diameter of the piston is 15 cm. What is the radius (in cm)? _____

    1 inch = 2.54 cm.

    What is the radius of the piston in inches? _____

    The area of the circular piston is found by $A = \pi r^2$.

    What is the area of the piston in square inches (in²)? _____

The pressure exerted on the piston by the added mass in pounds per square inch (psi) can be determined by dividing the mass in pounds by the area in square inches.

What is the pressure exerted by the added mass in psi? _____

The internal pressure is the sum of the external pressure and the added mass.

What is the calculated internal pressure? Compare your calculated answer with the internal pressure meter answer. How do they compare?

_____

6. *Predict the internal pressure (in psi) when 2.5 tons are added.* _____

How does your calculated answer compare with the internal pressure meter when you add 2.5 tons of mass? Record your data in the data table.

_____

6-1: Acid-Base Classification of Salts

In this assignment you will be asked to classify aqueous solutions of salts as to whether they are acidic, basic, or neutral. This is most easily done by first identifying how both the cation and anion affect the pH of the solution and then by combining the effects. After predicting the acid-base properties of these salts, you will then test your predictions in the laboratory.

1. *State whether 0.1 M solutions of each of the following salts are acidic, basic, or neutral. Explain your reasoning for each by writing a balanced net ionic equation to describe the behavior of each non-neutral salt in water: NaCN, KNO₃, NH₄Cl, NaHCO₃, and Na₃PO₄.*

   NaCN: _____
   _____
   _____

   KNO₃: _____
   _____

   NH₄Cl: _____
   _____
   _____

   NaHCO₃: _____
   _____
   _____

   Na₃PO₄: _____
   _____
   _____

Once you have predicted the nature of each salt solution, you will use *Virtual ChemLab* to confirm your prediction. Each solution must be approximately 0.1 M for your comparisons to be valid. Most of the solutions in the Stockroom are approximately 0.1 M already. Three solutions must be prepared from solid salts. One of these salt solutions is already prepared and on the lab bench ready for you to measure the pH.

2. Start *Virtual ChemLab,* select *Acid-Base Chemistry,* and then select *Acid-Base Classification of Salts* from the list of assignments. The lab will open in the Titrations laboratory.

3. On the stir plate, there will be a beaker of 0.10 M ammonium chloride (NH_4Cl) that has already been prepared. The pH meter has been calibrated and is in the beaker. Record the pH of the NH_4Cl solution in the data table on the following page. When finished, drag the beaker to the red disposal bucket, and drag the bottle of NH_4Cl to the stockroom counter.

4. Click in the *Stockroom* to enter. Double-click on the NH_4Cl bottle to return it to the shelf and then double-click on the $NaHCO_3$ and KNO_3 bottles to move them to the *Stockroom* counter. Return to the laboratory.

5. Open the beaker drawer (click on it) and drag a beaker to the spotlight next to the *Balance*. Click and drag the bottle of NaHCO₃ and place it on the spot light near the balance. Click in the *Balance* area to zoom in. Place a weigh paper on the balance and tare the balance. Open the bottle by clicking on the lid (*Remove Lid*). Pick up the *Scoop* and scoop up some salt by dragging the *Scoop* to the mouth of the bottle and then down the face of the bottle. Each scoop position on the face of the bottle represents a different size scoop. Pull the scoop down from the top to the second position (approximately 0.20 g) and drag it to the weigh paper in the balance until it snaps into place. Releasing the scoop places the sample on the weigh paper. Now drag the weigh paper from the balance to the beaker until it snaps into place and then empty the salt into the beaker. Return to the laboratory and drag the beaker to the stir plate.

6. Drag the 25 mL graduated cylinder to the sink under the tap until it fills. When filled, it will return to the lab bench and will indicate that it is full when you place the cursor over the cylinder. Drag the 25 mL cylinder to the beaker on the stir plate and empty it into the beaker. Place the pH probe in the beaker and record the pH in the data table. Drag the beaker to the red disposal bucket. Double-click the bottle of NaHCO₃ to move it to the *Stockroom* counter. Repeat steps 5 and 6 for KNO₃.

7. Click in the *Stockroom*. The stock solutions of NaCN and Na₃PO₄ are already approximately 0.1 M. Double-click each bottle to move them to the counter and return to the laboratory. With these solutions you can pour a small amount into a beaker that you have placed on the stir plate and place the pH probe in the solution. Record each pH in the data table. Drag each beaker to the red disposal bucket when you have finished. Were your predictions correct?

Data Table

solution	pH	acidic, basic or neutral
NH₄Cl		
NaHCO₃		
KNO₃		
NaCN		
Na₃PO₄		

6-2: Ranking Salt Solutions by pH

In this assignment you will be asked to rank aqueous solutions of acids, bases, and salts in order of increasing pH. This is most easily done by first identifying the strong acids that have the lowest pH, the strong bases that have the highest pH, and the neutral solutions that have a pH near 7. The weak acids will have a pH between 1 and 6 and the weak bases between 8 and 14. The exact order of weak acids and weak bases is determined by comparing the ionization constants (K_a for the weak acids and K_b for the weak bases). After ranking the pH of these solutions, you will then test your predictions in the laboratory.

1. *Arrange the following 0.1 M solutions in order of increasing pH and state why you placed each solution in that position: $NaCH_3COO$, HCl, HCN, NaOH, NH_3, NaCN, KNO_3, H_2SO_4, NH_4Cl, H_2SO_3, $NaHCO_3$, Na_3PO_4 and CH_3COOH.*

    In order of increasing pH:

Once you have predicted the nature of each salt solution, you will use *Virtual ChemLab* to confirm your prediction. Each solution must be approximately 0.1 M for your comparisons to be valid. Most of the solutions in the Stockroom are approximately 0.1 M already. Two solutions will need to be diluted and three solutions will need to be prepared from solid salts. One of these salt solutions is already prepared and on the lab bench ready for you to measure the pH.

2. Start *Virtual ChemLab*, select *Acid-Base Chemistry*, and then select *Ranking Salt Solutions by pH* from the list of assignments. The lab will open in the Titrations laboratory.

3. On the stir plate, there will be a beaker of 0.10 M ammonium chloride (NH_4Cl) that has already been prepared. The pH meter has been calibrated and is in the beaker. Record the pH of the NH_4Cl solution in the data table on the following page. When finished, drag the beaker to the red disposal bucket, and drag the bottle of NH_4Cl to the stockroom counter.

4. Click in the *Stockroom* to enter. Double-click on the NH₄Cl bottle to return it to the shelf and then double-click on the NaHCO₃ and KNO₃ bottles to move them to the *Stockroom* counter. Return to the laboratory

5. Open the beaker drawer (click on it) and drag a beaker to the spotlight next to the *Balance*. Click and drag the bottle of NaHCO₃ and place on the spot light near the balance. Click in the *Balance* area to zoom in. Place a weigh paper on the balance and tare the balance. Open the bottle by clicking on the lid (*Remove Lid*). Pick up the *Scoop* and scoop up some salt by dragging the *Scoop* to the bottle and then down the face of the bottle. Each scoop position on the face of the bottle represents a different size scoop. Pull the scoop down from the top to the second position (approximately 0.20 g) and drag it to the weigh paper in the balance until it snaps into place. Releasing the scoop places the sample on the weigh paper. Now drag the weigh paper from the balance to the beaker until it snaps into place and then empty the salt into the beaker. Return to the laboratory and drag the beaker to the stir plate.

6. Drag the 25 mL graduated cylinder to the sink under the tap until it fills. When filled, it will return to the lab bench and will indicate that it is full when you place the cursor over the cylinder. Drag the 25 mL cylinder to the beaker on the stir plate and empty it into the beaker. Place the pH probe in the beaker and record the pH in the data table. Drag the beaker to the red disposal bucket. Double-click the bottle of NaHCO₃ to move it to the *Stockroom* counter. Repeat steps 5 and 6 for KNO₃.

7. Click in the *Stockroom*. Double-click on the bottles of NH₃ and H₂SO₄ to move them from the shelf to the counter and return to the laboratory. Drag the bottle of NH₃ to one of the three spotlights on the lab bench. Place a beaker from the drawer on the stir plate. Drag the bottle of NH₃ to the 5 mL graduated cylinder (the smallest one) by the sink and fill the cylinder by dropping the bottle on the cylinder. Now drag the 5 mL graduated cylinder to the beaker on the stir plate and add the 5 mL of NH₃. Add 20 mL water to the beaker by filling and emptying the 10 mL cylinder into the beaker twice. Place the pH probe in the beaker and record the pH in the data table. Drag the beaker to the red disposal bucket. Double-click on the NH₃ bottle to move it back to the counter.

8. Repeat step 7 with H₂SO₄, *except* that you should use a 10 mL graduated cylinder of H₂SO₄ and adding 15 mL water.

9. Each of the other solutions is already approximately 0.1 M. With these solutions you can pour a small amount into the beaker that you have placed on the stir plate and place the pH probe in the solution to measure the pH. Record the pH of each solution in the data table. Drag each beaker to the red disposal bucket when you have finished. You must determine the pH for HCl, H₂SO₃, CH₃COOH (HAc), HCN, NaOH, NaCN, Na₃PO₄, and NaCH₃COO (NaAc). You may take two bottles at a time from the stockroom.

Data Table

solution	pH	solution	pH
NH₄Cl		CH₃COOH (HAc)	
NaHCO₃		HCN	
KNO₃		NaOH	
NH₃		NaCN	
H₂SO₄		NaCH₃COO (NaAc)	
HCl		Na₃PO₄	
H₂SO₃			

6-3: Concepts in Acid-Base Titrations

Titrations provide a method of quantitatively measuring the concentration of an unknown solution. In an acid-base titration, this is done by delivering a titrant of known concentration into an analyte of known volume. (The concentration of an unknown titrant can also be determined by titration with an analyte of known concentration and volume.) Titration curves (graphs of volume vs. pH) have characteristic shapes. The graph can be used to determine the strength or weakness of an acid or base. The equivalence point of the titration, or the point where the analyte has been completely consumed by the titrant, is identified by the point where the pH changes rapidly over a small volume of titrant delivered. There is a steep incline or decline at this point of the titration curve. It is also common to use an indicator that changes color at or near the equivalence point. In this assignment, you will observe this titration curve by titrating the strong acid HCl with the strong base NaOH.

1. Start *Virtual ChemLab*, select *Acid-Base Chemistry*, and then select *Concepts in Acid-Base Titrations* from the list of assignments. The lab will open in the Titration laboratory.

2. Click the *Lab Book* to open it. The buret will be filled with NaOH. The horizontal position of the orange handle is off for the stopcock. Click the **Save** button in the *Buret Zoom View* window. Open the stopcock by pulling down on the orange handle. The vertical position delivers solution the fastest with three intermediate rates in between. Turn the stopcock to one of the fastest positions. Observe the titration curve. When the volume reaches 35 mL, double-click the stopcock to stop the titration. Click **Stop** in the *Buret Zoom View*. A data link icon will be created in the lab book, click on it to view the data.

3. The beaker contains 0.3000 M HCl and the buret contains 0.3000 M NaOH.

4. *Write a complete balanced equation for the neutralization reaction between HCl and NaOH.*

_____

The following questions can be answered by examining the *Plot and Data Viewer* windows.

5. *What was the pH and color of the solution at the beginning of the titration?* _____

6. *What was the pH and color of the solution at the end of the titration?* _____

_____

7. Examine the graph of the pH vs. volume (blue line).

   Sketch the shape of the titration graph of pH vs. volume.

8. *What happens to the pH around 25 mL?* _____

9. *What would cause the change observed in question #4?* _____

   _____

   _____

   _____

6-4: Predicting the Equivalence Point

Titrations provide a method of quantitatively measuring the concentration of an unknown solution. In an acid-base titration, this is done by delivering a titrant of known concentration into an analyte of known volume. To make a titration more efficient and more accurate, it is often important to be able to predict the equivalence point for the titration. In this assignment, you will be given 0.3000 M HCl and 0.3000 M NaOH, you will predict the equivalence point, and then perform the titration to check your prediction.

1. Start *Virtual ChemLab,* select *Acid-Base Chemistry,* and then select *Predicting the Equivalence Point(1)* from the list of assignments. The lab will open in the Titrations laboratory.

2. Click the *Lab Book* to open it. Click the *Buret Zoom View* window to bring it to the front. The buret is filled with 0.3000 M NaOH. The beaker has 25.00 mL of 0.3000 M HCl. The pH meter is turned on and has been calibrated. The indicator is bromocresol green.

3. *Predict what volume (mL) of 0.3000 M NaOH is required to titrate the 25.00 mL of 0.3000 M HCl to the equivalence point.*

4. *Perform the titration.* Click the **Save** button in the *Buret Zoom View* window so the titration data can be saved. The horizontal position of the orange handle is off for the stopcock. Open the stopcock by pulling down on the orange handle. The vertical position delivers solution the fastest with three intermediate rates in between. Turn the stopcock to one of the fastest positions. Observe the titration curve. When the blue line in the graph window (the pH curve) begins to turn up, double-click the stopcock to turn it off. Move the stopcock down one position to add volume drop by drop.

   There are two methods for determining the volume at the equivalence point: (1) Stop the titration (close the stopcock) when a color change occurs, and then click the **Stop** button in the *Buret Zoom View*. A blue data link will appear in the lab book. Click the data link icon to open the *Data View* window. Scroll down to the last data entry and record the volume at the equivalence point. OR (2) Add drops slowly through the equivalence point until the pH reaches approximately 12. Click the **Stop** button in the *Buret Zoom View*. A data link icon will appear in the lab book. Click the data link icon to open the *Data View* window. Click **Copy Data** button to copy and paste the data to a spreadsheet program. Plot the first derivative of pH vs. volume. The peak will indicate the volume of the equivalence point since this is where the pH is changing the most rapidly as the volume changes.

5. *What volume of 0.3000 M NaOH was required by the titration to reach the equivalence point?*

   _____

6. *Calculate the percent error of the predicted volume using the formula:*

$$\% \ Error = \frac{|your\ predicted\ answer - your\ actual\ answer|}{your\ predicted\ answer} \times 100$$

% Error =

If you want to repeat the titration, click *Exit*, select this problem from the workbook again, and repeat the experiment.

6-5: Predicting the Equivalence Point

Titrations provide a method of quantitatively measuring the concentration of an unknown solution. In an acid-base titration, this is done by delivering a titrant of known concentration into an analyte of known volume. To make a titration more efficient and more accurate, it is often important to be able to predict the equivalence point for the titration. In this assignment, you will be given 0.1276 M HCl and 0.1475 M NaOH, you will predict the equivalence point, and then perform the titration to check your prediction.

1. Start *Virtual ChemLab,* select *Acid-Base Chemistry,* and then select *Predicting the Equivalence Point(2)* from the list of assignments. The lab will open in the Titrations laboratory.

2. Click the *Lab Book* to open it. Click the *Buret Zoom View* window to bring it to the front. The buret is filled with 0.1475 M NaOH. The beaker has 25.00 mL of 0.1276 M HCl. The pH meter is turned on and has been calibrated. The indicator is bromocresol green.

3. *Predict what volume (mL) of 0.1475 M NaOH is required to titrate the 25.00 mL of 0.1276 M HCl to the equivalence point.*

4. *Perform the titration.* Click the **Save** button in the *Buret Zoom View* window so the titration data can be saved. The horizontal position of the orange handle is off for the stopcock. Open the stopcock by pulling down on the orange handle. The vertical position delivers solution the fastest with three intermediate rates in between. Turn the stopcock to one of the fastest positions. Observe the titration curve. When the blue line in the graph window (the pH curve) begins to turn up, double-click the stopcock to turn it off. Move the stopcock down one position to add volume drop by drop.

There are two methods for determining the volume at the equivalence point: (1) Stop the titration (close the stopcock) when a color change occurs, and then click the **Stop** button in the *Buret Zoom View.* A blue data link will appear in the lab book. Click the data link icon to open the *Data View* window. Scroll down to the last data entry and record the volume at the equivalence point. OR (2) Add drops slowly through the equivalence point until the pH reaches approximately 12. Click the **Stop** button in the *Buret Zoom View.* A data link icon will appear in the lab book. Click the data link icon to open the *Data View* window. Click **Select All** button to copy and paste the data to a spreadsheet program. Plot the first derivative of pH vs. volume. The peak will indicate the volume of the equivalence point since this is where the pH is changing the most rapidly as the volume changes.

5. *What volume of 0.1475 M NaOH was required by the titration to reach the equivalence point?*

_____

6. *Calculate the percent error of the predicted volume using the formula:*

$$\% \, Error = \frac{\left| your \; predicted \; answer - your \; actual \; answer \right|}{your \; predicted \; answer} \times 100$$

% Error =

If you want to repeat the titration, click *Exit*, select this problem from the workbook again, and repeat the experiment.

6-6: Predicting the Equivalence Point

Titrations provide a method of quantitatively measuring the concentration of an unknown solution. In an acid-base titration, this is done by delivering a titrant of known concentration into an analyte of known volume. To make a titration more efficient and more accurate, it is often important to be able to predict the equivalence point for the titration. In this assignment, you will be given 0.1033 M CH_3COOH (acetic acid or HAc) and 0.1104 M NaOH, you will predict the equivalence point, and then perform the titration to check your prediction.

1. Start *Virtual ChemLab*, select *Acid-Base Chemistry*, and then select *Predicting the Equivalence Point(3)* from the list of assignments. The lab will open in the Titrations laboratory.

2. Click the *Lab Book* to open it. Click the *Buret Zoom View* window to bring it to the front. The buret is filled with 0.1104 M NaOH. The beaker has 15.00 mL of 0.1033 M HAc. The pH meter is turned on and has been calibrated. The indicator is methyl orange.

3. *Predict what volume (mL) of 0.1104 M NaOH is required to titrate the 15.00 mL of 0.1033 M HAc to the equivalence point.*

4. *Perform the titration.* Click the **Save** button in the *Buret Zoom View* window so the titration data can be saved. The horizontal position of the orange handle is off for the stopcock. Open the stopcock by pulling down on the orange handle. The vertical position delivers solution the fastest with three intermediate rates in between. Turn the stopcock to one of the fastest positions. Observe the titration curve. When the blue line in the graph window (the pH curve) begins to turn up, double-click the stopcock to turn it off. Move the stopcock down one position to add volume drop by drop.

    There are two methods for determining the volume at the equivalence point: (1) Stop the titration (close the stopcock) when a color change occurs, and then click the **Stop** button in the *Buret Zoom View*. A blue data link will appear in the lab book. Click the data link icon to open the *Data View* window. Scroll down to the last data entry and record the volume at the equivalence point. OR (2) Add drops slowly through the equivalence point until the pH reaches approximately 12. Click the **Stop** button in the *Buret Zoom View*. A data link icon will appear in the lab book. Click the data link icon to open the *Data View* window. Click **Select All** button to copy and paste the data to a spreadsheet program. Plot the first derivative of pH vs. volume. The peak will indicate the volume of the equivalence point since this is where the pH is changing the most rapidly as the volume changes.

5. *What volume of 0.1104 M NaOH was required by the titration to reach the equivalence point?*

_____

6. *Calculate the percent error of the predicted volume using the formula:*

$$\% \ Error = \frac{|your \ predicted \ answer - your \ actual \ answer|}{your \ predicted \ answer} \times 100$$

% Error =

If you want to repeat the titration, click *Exit*, select this problem from the workbook again, and repeat the experiment.

6-7: Ionization Constants of Weak Acids

An acid-base indicator is usually a weak acid with a characteristic color in the protonated and deprotonated forms. In this assignment, you will monitor the color of an acetic acid solution containing *Bromocresol Green* as an indicator, as the pH is changed and then you will estimate the ionization constant, K_a, for the indicator.

1. Start *Virtual ChemLab,* select *Acid-Base Chemistry,* and then select *Ionization Constants of Weak Acids* from the list of assignments. The lab will open in the Titration laboratory. Bottles of 0.1104 M NaOH and 0.1031 M HAc (acetic acid) will be on the lab bench. The buret will be filled with the NaOH solution and a beaker containing 10.00 mL of the HAc solution will be on the stir plate. The stir plate will be on, *Bromocresol Green* indicator will have been added to the beaker, and a calibrated pH probe will also be in the beaker so the pH of the solution can be monitored.

2. *What is the color and pH of the solution?* _____

3. On the buret, the horizontal position of the orange handle is off for the stopcock. Open the stopcock by pulling down on the orange handle. The vertical position delivers solution the fastest with three intermediate rates in between (slow drop-wise, fast drop-wise, and slow stream). Turn the stopcock to the second position or fast drop-wise addition. Observe the color of the solution and close the stopcock when the color turns green by double clicking on the center of the stopcock.

4. *What is the color and pH of the solution now?* _____

5. Continue to add NaOH as before or at a faster rate.

   What is the final color of the solution? _____

6. An acid-base indicator is usually a weak acid with a characteristic color in the protonated and deprotonated forms. Because bromocresol green is an acid, it is convenient to represent its rather complex formula as HBCG. HBCG ionizes in water according to the following equation:

$$HBCG + H_2O = BCG^- + H_3O^+$$
(yellow)   (blue)

The K_a (the equilibrium constant for the acid) expression is:

$$K_a = \frac{[BCG^-][H_3O^+]}{[HBCG]}$$

When [BCG$^-$] = [HBCG], then K_a = [H$_3$O$^+$]. If you know the pH of the solution, then the [H$_3$O$^+$] and K_a can be determined.

What would be the color of the solution if there were equal concentrations of HBCG and BCG$^-$?

_____

7. *What is the pH at the first appearance of this color?* _____

_____

8. *What is an estimate for the K_a for bromocresol green?* _____

_____

6-8: Acid-Base Titration: Practice

Titrations provide a method of quantitatively measuring the concentration of an unknown solution. In an acid-base titration, this is done by delivering a titrant of known concentration into an analyte of known volume. In this assignment, you will titrate a 0.3000 M solution of NaOH into 25 mL of 0.3000 M HCl. Although in this case you know the concentration of both NaOH and HCl, this will give you practice in performing a titration and calculating the concentration of the analyte, which in this case is HCl.

1. Start *Virtual ChemLab,* select *Acid-Base Chemistry,* and then select *Acid-Base Titration: Practice* from the list of assignments. The lab will open in the Titrations laboratory.

2. Click the *Lab Book* to open it. Click the *Buret Zoom View* window to bring it to the front. The buret is filled with 0.3000 M NaOH. The beaker has 25.00 mL of 0.3000 M HCl. The pH meter is turned on and has been calibrated. The indicator is bromocresol green.

3. *Perform the titration.* Click the *Save* button in the *Buret Zoom View* window so the titration data can be saved. The horizontal position of the orange handle is off for the stopcock. Open the stopcock by pulling down on the orange handle. The vertical position delivers solution the fastest with three intermediate rates in between. Turn the stopcock to one of the fastest positions. Observe the titration curve. When the blue line in the graph window (the pH curve) begins to turn up, double-click the stopcock to turn it off. Move the stopcock down one position to add volume drop by drop.

    There are two methods for determining the volume at the equivalence point: (1) Stop the titration (close the stopcock) when a color change occurs, and then click the *Stop* button in the *Buret Zoom View*. A blue data link will appear in the lab book. Click the data link icon to open the *Data View* window. Scroll down to the last data entry and record the volume at the equivalence point. OR (2) Add drops slowly through the equivalence point until the pH reaches approximately 12. Click the *Stop* button in the *Buret Zoom View*. A data link icon will appear in the lab book. Click the data link icon to open the *Data View* window. Click *Select All* button to copy and paste the data to a spreadsheet program. Plot the first derivative of pH vs. volume. The peak will indicate the volume of the equivalence point since this is where the pH is changing the most rapidly as the volume changes.

4. *What volume of 0.3000 M NaOH was required by the titration to reach the equivalence point?*

    _____

5. *Calculate the molarity of the HCl using 25.00 mL of HCl solution and the volume of the 0.3000 M NaOH from your titration.*

6. *Remember that the concentration of your HCl solution is 0.3000 M. Calculate the percent error using the formula:*

$$\% \ Error = \frac{|your \ calculated \ answer - the \ actual \ answer|}{the \ predicted \ answer} \times 100$$

% Error =

If you want to repeat the titration, click *Exit*, select this problem from the workbook again, and repeat the experiment.

6-9: Acid-Base Titration: Unknown HCl

Titrations provide a method of quantitatively measuring the concentration of an unknown solution. In an acid-base titration, this is done by delivering a titrant of known concentration into an analyte of known volume. In this assignment, you will titrate a 0.2564 M solution of NaOH into 25 mL of an unknown concentration of HCl and calculate the concentration of the HCl solution.

1. Start *Virtual ChemLab,* select *Acid-Base Chemistry,* and then select *Acid-Base Titration: Unknown HCl* from the list of assignments. The lab will open in the Titrations laboratory.

2. Click the *Lab Book* to open it. Click the *Buret Zoom View* window to bring it to the front. The buret is filled with 0.2564 M NaOH. The beaker has 25.00 mL of unknown HCl. The pH meter is turned on and has been calibrated. The indicator is bromocresol green.

3. *Perform the titration.* Click the **Save** button in the *Buret Zoom View* window so the titration data can be saved. The horizontal position of the orange handle is off for the stopcock. Open the stopcock by pulling down on the orange handle. The vertical position delivers solution the fastest with three intermediate rates in between. Turn the stopcock to one of the fastest positions. Observe the titration curve. When the blue line in the graph window (the pH curve) begins to turn up, double-click the stopcock to turn it off. Move the stopcock down one position to add volume drop by drop.

   There are two methods for determining the volume at the equivalence point: (1) Stop the titration (close the stopcock) when a color change occurs, and then click the **Stop** button in the *Buret Zoom View*. A data link icon will appear in the lab book. Click the data link icon to open the *Data View* window. Scroll down to the last data entry and record the volume at the equivalence point. OR (2) Add drops slowly through the equivalence point until the pH reaches approximately 12. Click the **Stop** button in the *Buret Zoom View*. A data link icon will appear in the lab book. Click the data link icon to open the *Data View* window. Click **Copy Data** button to copy and paste the data to a spreadsheet program. Plot the first derivative of pH vs. volume. The peak will indicate the volume of the equivalence point since this is where the pH is changing the most rapidly as the volume changes.

4. *What is your unknown sample number?* _____

   _____

5. *What volume of 0.2564 M NaOH was required by the titration to reach the equivalence point?*

   _____

   _____

6. *Calculate the molarity of the HCl using 25.00 mL of HCl solution and the volume of the 0.2564 M*

   NaOH from your titration. _____

6-10: Study of Acid-Base Titrations – Monoprotic Acids

Titrations provide a method of quantitatively measuring the concentration of an unknown solution. In an acid-base titration, this is done by delivering a titrant of known concentration into an analyte of known volume. (The concentration of an unknown titrant can also be determined by titration with an analyte of known concentration and volume.) Titration curves (graphs of volume vs. pH) have characteristic shapes. The graph can be used to determine the strength or weakness of an acid or base. The equivalence point of the titration, or the point where the analyte has been completely consumed by the titrant, is identified by the point where the pH changes rapidly over a small volume of titrant delivered. In this assignment, you will observe this titration curve by titrating the strong acid HCl with the strong base NaOH.

1. Start *Virtual ChemLab,* select *Acid-Base Chemistry,* and then select *Study of Acid-Base Titrations – Monoprotic Acids* from the list of assignments. The lab will open in the Titration laboratory.

2. Click the *Lab Book* to open it. The buret will be filled with NaOH and 25.00 mL of HCl will be in the beaker with bromocresol green as an indicator. Click the **Save** button in the *Buret Zoom View* window. The horizontal position of the orange handle is off for the stopcock. Open the stopcock by pulling down on the orange handle. The vertical position delivers solution the fastest with three intermediate rates in between. Turn the stopcock to one of the fastest positions. Observe the titration curve. When the volume reaches 35 mL, double-click the stopcock to stop the titration. Click **Stop** in the *Buret Zoom View*. A data link icon will be created in the lab book. Click on it to view the titration data.

    If you need to repeat the titration, click in the *Stockroom* to enter, click on the clipboard, and select Preset Experiment #1 *Strong Acid-Strong Base.*

3. The beaker contains 0.3000 M HCl and the buret contains 0.3000 M NaOH.

    Write a complete balanced equation for the neutralization reaction between HCl and NaOH.

    _____

    The following questions can be answered by examining the *Plot* window and the *Data Viewer* window.

4. *What was the pH and color of the solution at the beginning of the titration?* _____

5. *What was the pH and color of the solution at the end of the titration?*

    _____

6. *Examine the graph of pH vs. volume (blue line) and sketch the titration curve below.*

7. *What happens to the pH around 25 mL and what causes this?*

_____

_____

_____

_____

8. *Examine the graph of conductivity vs. volume (red line) and sketch the titration curve on the graph above.*

9. *What happens to the conductivity during the titration?* _____

_____

10. *What would cause the change observed in question #9?*

_____

_____

_____

_____

6-11: Weak Acid-Strong Base Titrations

Titrations provide a method of quantitatively measuring the concentration of an unknown solution. In an acid-base titration, this is done by delivering a titrant of known concentration into an analyte of known volume. Titration curves (graphs of volume vs. pH) have characteristic shapes. The equivalence point of the titration, or the point where the analyte has been completely consumed by the titrant, is identified by the point where the pH changes rapidly over a small volume of titrant delivered. In this assignment, you will observe this titration curve by titrating the weak acid CH_3COOH (acetic acid) with the strong base NaOH. You will also predict the pH at the equivalence point, validate your prediction experimentally, and then calculate the equilibrium constant for the neutralization reaction.

1. Start *Virtual ChemLab, select Acid-Base Chemistry,* and then select *Weak Acid-Strong Base Titrations* from the list of assignments. The lab will open in the Titration laboratory.

2. Click the *Lab Book* to open it. The buret will be filled with NaOH and 25.00 mL of CH_3COOH will be in the beaker with phenolphthalein as the indicator. Click the *Save* button in the *Buret Zoom View* window. The horizontal position of the orange handle is off for the stopcock. Open the stopcock by pulling down on the orange handle. The vertical position delivers solution the fastest with three intermediate rates in between. Turn the stopcock to one of the fastest positions. Observe the titration curve. When the volume reaches 40 mL, double-click the stopcock to stop the titration. Click *Stop* in the *Buret Zoom View*. A data link icon will be created in the lab book. Click on it to view the data.

    If you need to repeat the titration, click in the *Stockroom* to enter, click on the clipboard, and select Preset Experiment # 3 *Weak Acid-Strong Base.*

2. The beaker contains 0.1894 M CH_3COOH and the buret contains 0.2006 M NaOH.

    Write a complete balanced equation for the neutralization reaction between CH_3COOH and NaOH and then write a balanced net ionic equation for this chemical reaction.

    The following questions can be answered by examining the *Plot* window, the *Data Viewer* window, and the balanced net ionic equation.

4. *Indicate the species present at the equivalence point, and predict whether the pH at the equivalence point will be pH > 7, pH < 7, or pH ≈ 7. Explain why you made this prediction.*

    _____

    _____

5. *Examine the graph of pH vs. volume (blue line) and sketch the titration curve on the following page. Mark the equivalence point as halfway between the top and bottom "shoulders" of the curve. Based on your graph, what is the pH at the equivalence point? How does it compare with your predicted pH?*

6. *Calculate the value of the equilibrium constant for the balanced net ionic equation for the weak acid-strong base titration.*

6-12: Strong Acid-Weak Base Titrations

Titrations provide a method of quantitatively measuring the concentration of an unknown solution. In an acid-base titration, this is done by delivering a titrant of known concentration into an analyte of known volume. Titration curves (graphs of volume vs. pH) have characteristic shapes. The equivalence point of the titration, or the point where the analyte has been completely consumed by the titrant, is identified by the point where the pH changes rapidly over a small volume of titrant delivered. In this assignment, you will observe this titration curve by titrating the strong acid HCl with the weak base sodium hydrogen carbonate ($NaHCO_3$). You will also predict the pH at the equivalence point, validate your prediction experimentally, and then calculate the equilibrium constant for the neutralization reaction.

1. Start *Virtual ChemLab,* select *Acid-Base Chemistry,* and then select *Strong Acid-Weak Base Titrations* from the list of assignments. The lab will open in the Titration laboratory.

2. Click the *Lab Book* to open it. The buret will be filled with HCl and 25 mL of $NaHCO_3$ solution will be in the beaker with methyl orange as the indicator. Click the **Save** button in the *Buret Zoom View* window. The horizontal position of the orange handle is off for the stopcock. Open the stopcock by pulling down on the orange handle. The vertical position delivers solution the fastest with three intermediate rates in between. Turn the stopcock to one of the fastest positions. Observe the titration curve. When the volume reaches 40 mL, double-click the stopcock to stop the titration. Click **Stop** in the *Buret Zoom View*. A data link icon will be created in the lab book. Click on it to view the data.

    If you need to repeat the titration, click in the *Stockroom* to enter, click on the clipboard, and select Preset Experiment # 5 *Strong Acid-Weak Base.*

3. The beaker contains 0.40 M $NaHCO_3$ and the buret contains 0.30 M HCl.

    Write a complete balanced equation for the neutralization reaction between HCl and $NaHCO_3$ and then write a balanced net ionic equation for this chemical reaction.

    The following questions can be answered by examining the *Plot* window, the *Data Viewer* window, and the balanced net ionic equation.

4. *Indicate the species present at the equivalence point, and predict whether the pH at the equivalence point will be pH > 7, pH < 7, or pH ≈ 7. Explain why you made this prediction.*

    _____

    _____

5. *Examine the graph of pH vs. volume (blue line) and sketch the titration curve on the following page. Mark the equivalence point as halfway between the top and bottom "shoulders" of the curve. Based on your graph, what is the pH at the equivalence point? How does it compare with your predicted pH?*

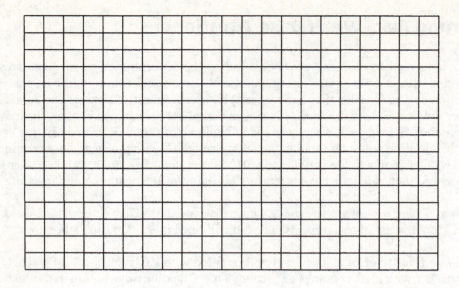

6. *Calculate the value of the equilibrium constant for the balanced net ionic equation for the strong acid-weak base titration.*

_____

_____

_____

6-13: Weak Acid-Weak Base Titrations

Titrations provide a method of quantitatively measuring the concentration of an unknown solution. In an acid-base titration, this is done by delivering a titrant of known concentration into an analyte of known volume. Titration curves (graphs of volume vs. pH) have characteristic shapes. The equivalence point of the titration, or the point where the analyte has been completely consumed by the titrant, is identified by the point where the pH changes rapidly over a small volume of titrant delivered. In this assignment, you will observe this titration curve by titrating the weak acid CH_3COOH (acetic acid) with the weak base NH_3. You will also predict the pH at the equivalence point, validate your prediction experimentally, and then calculate the equilibrium constant for the neutralization reaction.

1. Start *Virtual ChemLab,* select *Acid-Base Chemistry,* and then select *Weak Acid-Weak Base Titrations* from the list of assignments. The lab will open in the Titration laboratory.

2. Click the *Lab Book* to open it. The buret will be filled with NH_3 and 50.00 mL of CH_3COOH is in the beaker with phenolphthalein as the indicator. Click the **Save** button in the *Buret Zoom View* window. The horizontal position of the orange handle is off for the stopcock. Open the stopcock by pulling down on the orange handle. The vertical position delivers solution the fastest with three intermediate rates in between. Turn the stopcock to one of the fastest positions. Observe the titration curve. When the volume reaches 25 mL, double-click the stopcock to stop the titration. Click **Stop** in the *Buret Zoom View*. A data link icon will be created in the lab book. Click on it to view the data.

   If you need to repeat the titration, exit the laboratory and select *Weak Acid-Weak Base Titrations* from the list of assignments.

3. The beaker contains 0.1033 M CH_3COOH and the buret contains 0.4949 M NH_3.

   Write a complete balanced equation for the neutralization reaction between CH_3COOH and NH_3.

   The following questions can be answered by examining the *Plot* window, the *Data Viewer* window, and the balanced equation.

4. *Indicate the species present at the equivalence point, and predict whether the pH at the equivalence point will be pH > 7, pH < 7, or pH ≈ 7. Explain why you made this prediction.*

   _____

   _____

   _____

5. *Examine the graph of pH vs. volume (blue line) and sketch the titration curve on the following page. Mark the equivalence point as halfway between the top and bottom "shoulders" of the curve. Based on your graph, what is the pH at the equivalence point? How does it compare with your predicted pH?*

6. *Calculate the value of the equilibrium constant for the balanced net ionic equation for the weak acid-weak base titration.*

_____

_____

_____

_____

6-14: Study of Acid-Base Titrations – Polyprotic Acids

Titrations provide a method of quantitatively measuring the concentration of an unknown solution. In an acid-base titration, this is done by delivering a titrant of known concentration into an analyte of known volume. (The concentration of an unknown titrant can also be determined by titration with an analyte of known concentration and volume.) Titration curves (graphs of volume vs. pH) have characteristic shapes. The graph can be used to determine the strength or weakness of an acid or base. The equivalence point of the titration, or the point where the analyte has been completely consumed by the titrant, is identified by the point where the pH changes rapidly over a small volume of titrant delivered. For polyprotic acids, there will be multiple equivalence points. In this assignment, you will observe this titration curve by titrating the weak acid H_2SO_3 with the strong base NaOH.

1. Start *Virtual ChemLab,* select *Acid-Base Chemistry,* and then select *Study of Acid-Base Titrations – Polyprotic Acids* from the list of assignments. The lab will open in the Titration laboratory.

2. Click the *Lab Book* to open it. The buret will be filled with NaOH and 25.00 mL of H_2SO_3 will be in the beaker with thymol blue as an indicator. Click the ***Save*** button in the *Buret Zoom View* window. The horizontal position of the orange handle is off for the stopcock. Open the stopcock by pulling down on the orange handle. The vertical position delivers solution the fastest with three intermediate rates in between. Turn the stopcock to one of the fastest positions. Observe the titration curve. When the volume reaches 40 mL, double-click the stopcock to stop the titration. Click ***Stop*** in the *Buret Zoom View*. A data link icon will be created in the lab book. Click on it to view the titration data.

   If you need to repeat the titration, click in the *Stockroom* to enter, click on the clipboard and select Preset Experiment #7 *Polyprotic Acid-Strong Base*.

3. The beaker contains 0.2556 M H_2SO_3 and the buret contains 0.3106 M NaOH.

   Write a complete balanced equation for the two-step neutralization reaction between H_2SO_3 and NaOH.

   The following questions can be answered by examining the *Plot* window and the *Data Viewer* window.

4. *What was the pH and color of the solution at the beginning of the titration?* _____

5. *What was the pH and color of the solution at the end of the titration? Did any additional color changes occur during the titration?*

   _____

6. *Examine the graph of pH vs. volume (blue line) and sketch the titration curve on the following page.*

7. *What happens to the pH around 16 mL and 32 mL? What causes each to occur?*

_____

_____

_____

_____

_____

_____

_____

8. *Examine the graph of conductivity vs. volume (red line) and sketch the titration curve on the graph above.*

9. *What happens to the conductivity during the titration?*

_____

_____

10. *What would cause the change observed in question #10?*

_____

_____

_____

_____

_____

_____

6-15: Acid-Base Standardization

Titrations provide a method of quantitatively measuring the concentration of an unknown solution. In an acid-base titration, this is done by delivering a titrant of known concentration into an analyte of known volume. (The concentration of an unknown titrant can also be determined by titration with an analyte of known concentration and volume.) Titration curves (graphs of volume vs. pH) have characteristic shapes. The graph can be used to determine the strength or weakness of an acid or base. The equivalence point of the titration, or the point where the analyte has been completely consumed by the titrant, is identified by the point where the pH changes rapidly over a small volume of titrant delivered. In this assignment, you will determine the molarity of an unknown solution of NaOH by titrating against a primary standard, potassium hydrogen phthalate (KHP).

1. Start *Virtual ChemLab,* select *Acid-Base Chemistry,* and then select *Acid-Base Standardization* from the list of assignments. The lab will open in the Titrations laboratory.

2. Click the *Lab Book* to open it. Click the *Beakers* drawer and place a beaker in the spotlight next to the balance. Click on the *Balance* area to zoom in, open the bottle of KHP by clicking on the lid (*Remove Lid*). Drag the beaker to the balance to place it on the balance pan and tare the balance. Pick up the *Scoop* and scoop out some sample by first dragging the scoop to the mouth of the bottle and then pulling the scoop down the face of the bottle. As the scoop is dragged down the face of the bottle it will pickup different quantities of solid. Select the largest sample possible and drag the scoop to the beaker on the balance until it snaps in place and then let go. Repeat this one additional time so you have put two scoops (approximately 2 g) of KHP in the beaker. Record the mass of the sample in the data table on the following page and return to the laboratory.

3. Drag the beaker from the balance to the sink and hold it under the tap to add a small amount of water. Place it on the stir plate and drag the calibrated pH meter probe to the beaker. Add *Phenolphthalein* as the indicator.

4. The buret will be filled with NaOH. Click the *Save* button in the *Buret Zoom View* window so the titration data can be saved. The horizontal position of the orange handle is off for the stopcock. Open the stopcock by pulling down on the orange handle. The vertical position delivers solution the fastest with three intermediate rates in between. Turn the stopcock to one of the fastest positions. Observe the titration curve. When the blue line begins to turn up, double-click the stopcock to turn it off. Move the stopcock down one position to add volume drop by drop.

    There are two methods for determining the volume at the equivalence point: (1) Stop the titration when a color change occurs. Click the *Stop* button in the *Buret Zoom View.* A data link icon will appear in the lab book. Click the data link icon to open the *Data Viewer* window. Scroll down to the last data entry and record the volume at the equivalence point in the data table. OR (2) Add drops slowly through the equivalence point until the pH reaches approximately 12. Click the *Stop* button in the *Buret Zoom View.* A data link icon will appear in the lab book. Click the data link icon to open the *Data Viewer* window. Click the *Copy Data* button to copy and paste the data to a spreadsheet. Plot the first derivative of pH vs. volume. The peak will indicate the volume at the equivalence point since this is where the pH is changing the most rapidly as the volume changes.

    Repeat at least two additional times recording data in the data table. Do not forget to refill the buret with NaOH and place the pH meter and indicator in the beaker each time.

    The molecular weight of KHP is 204.22 g/mol.

Unknown # _____

Data Table

Trial	mass KHP (g)	volume NaOH (mL)	molarity NaOH (mol/L)
1			
2			
3			
4			
5			

5. *Write a balanced chemical equation for the reaction of KHP and NaOH.*

    _____

6. *What is the average molarity of the unknown NaOH for your closest three titrations?*

    _____

6-16: Analysis of Baking Soda

Titrations provide a method of quantitatively measuring the concentration of an unknown solution. In an acid-base titration, this is done by delivering a titrant of known concentration into an analyte of known volume. (The concentration of an unknown titrant can also be determined by titration with an analyte of known concentration and volume.) Titration curves (graphs of volume vs. pH) have characteristic shapes. The graph can be used to determine the strength or weakness of an acid or base. The equivalence point of the titration, or the point where the analyte has been completely consumed by the titrant, is identified by the point where the pH changes rapidly over a small volume of titrant delivered. In this assignment, you will determine the mass % of an unknown sample of baking soda ($NaHCO_3$) by titrating it with an HCl solution of known concentration.

1. Start *Virtual ChemLab,* select *Acid-Base Chemistry,* and then select *Analysis of Baking Soda* from the list of assignments. The lab will open in the Titration laboratory. The laboratory will open with a beaker on the stir plate with 1.5000 g of impure solid $NaHCO_3$ and with sufficient water added to make the total volume 25.00 mL. Methyl orange indicator will have also been added to the beaker, as well as the calibrated pH meter probe.

2. Click on the *Lab Book* to open it and click on the *Buret Zoom View* window and the pH meter window to bring them to the front. The buret will be filled with 0.3015 M HCl. Click the *Save* button in the *Buret Zoom View* window so the titration data can be saved. The horizontal position of the orange handle is off for the stopcock. Open the stopcock by pulling down on the orange handle. The vertical position delivers solution the fastest with three intermediate rates in between. Turn the stopcock to one of the fastest positions. Observe the titration curve. When the blue line begins to turn down, double-click the stopcock to turn it off. Move the stopcock down one position to add volume drop by drop.

   There are two methods for determining the volume at the equivalence point: (1) Stop the titration when a color change occurs. Click the *Stop* button in the *Buret Zoom View*. A data link icon will appear in the lab book. Click the data link icon to open the *Data Viewer* window. Scroll down to the last data entry and record the volume at the equivalence point in the data table below. OR (2) Add drops slowly through the equivalence point until the pH reaches approximately 2. Click the *Stop* button in the *Buret Zoom View*. A data link icon will appear in the lab book. Click the data link icon to open the *Data Viewer* window. Click the *Copy Data* button to copy and paste the data to a spreadsheet. Plot the first derivative of pH vs. volume. The peak will indicate the volume at the equivalence point since this is where the pH is changing the most rapidly as the volume changes.

   Unknown sample # _____

Data Table

mass unknown sample (g)	volume HCl (mL)	molarity HCl (mol/L)

3. *Write a balanced chemical equation for the reaction between NaHCO₃ and HCl.*

   _____

4. *Calculate the moles of HCl by multiplying the volume of HCl in liters and the molarity of HCl in*

mol/L. *(Keep four significant digits in all of the calculations.)* _____

5. *The moles of HCl can be converted to moles of $NaHCO_3$ using the coefficients from the balanced equation. What is the mole to mole ratio of HCl to $NaHCO_3$? How many moles of $NaHCO_3$ are present in the sample?*

_____

6. *Calculate the grams of $NaHCO_3$ by multiplying the moles of $NaHCO_3$ by the molecular weight of $NaHCO_3$ (84.007 g/mol).*

_____

7. The mass % of $NaHCO_3$ present in the sample can be calculated by dividing the mass of $NaHCO_3$ from question #6 by the mass of the sample from the data table and multiplying by 100.

What is the mass % of $NaHCO_3$?

_____

7-1: Study of Oxidation-Reduction Titrations

Titrations provide a method of quantitatively measuring the concentration of an unknown solution. This is done by delivering a titrant of known concentration into an analyte of known volume. (The concentration of an unknown titrant can also be determined by titration with an analyte of known concentration and volume.) In oxidation-reduction (redox) titrations, the voltage resulting from the mixture of an oxidant and reductant can be measured as the titration proceeds. The equivalence point of the titration, or the point where the analyte has been completely consumed by the titrant, is identified by the point where the voltage changes rapidly over a small volume of titrant delivered. In this assignment, you will observe this titration curve by titrating $FeCl_2$ with $KMnO_4$.

1. Start *Virtual ChemLab,* select *Electrochemistry,* and then select *Study of Oxidation-Reduction Titrations* from the list of assignments. The lab will open in the Titrations laboratory.

2. The buret will be filled with $KMnO_4$ and a solution containing $FeCl_2$ will be in the beaker on the stir plate. The horizontal position of the orange handle is off for the stopcock. Open the stopcock by pulling down on the orange handle. The vertical position delivers volume the fastest with three intermediate rates in between. Turn the stopcock to one of the fastest positions. Observe the titration curve. When the volume reaches 45 mL double-click the stopcock to turn it off.

3. Examine the graph of voltage vs. volume (blue line) and sketch the titration curve below. Label the axes.

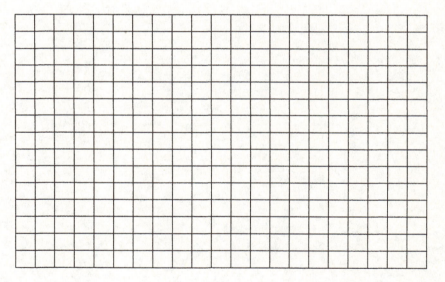

4. *Write a balanced net ionic equation for the reaction in acidic solution of $FeCl_2$ and $KMnO_4$ (Fe^{2+} becomes Fe^{3+} and MnO_4^- becomes Mn^{2+}).*

5. The reduction potential of Fe^{2+} is 0.732 volts, and the reduction potential of MnO_4^- in acidic solution is 1.507 volts.

    If you titrate $KMnO_4$ into $FeCl_2$, what happens to the voltage of the solution as the titration starts and proceeds to the end?

7-2: Standardization of a Permanganate Solution

Titrations provide a method of quantitatively measuring the concentration of an unknown solution. This is done by delivering a titrant of known concentration into an analyte of known volume. (The concentration of an unknown titrant can also be determined by titration with an analyte of known concentration and volume.) In oxidation-reduction (redox) titrations, the voltage resulting from the mixture of an oxidant and reductant can be measured as the titration proceeds. The equivalence point of the titration, or the point where the analyte has been completely consumed by the titrant, is identified by the point where the voltage changes rapidly over a small volume of titrant delivered. In this assignment, you will determine the molarity of an unknown solution of $KMnO_4$ by titrating against a primary standard, solid As_2O_3.

1. Start *Virtual ChemLab,* select *Electrochemistry,* and then select *Standardization of a Permanganate Solution* from the list of assignments. The lab will open in the Titrations laboratory.

2. Click the *Lab Book* to open it. Click the *Beakers* drawer and place a beaker in the spotlight next to the balance. Click on the *Balance* area to zoom in, open the bottle of As_2O_3 by clicking on the lid (*Remove Lid*). Drag the beaker to the balance to place it on the balance pan and tare the balance. Pick up the *Scoop* and scoop out some sample by first dragging the scoop to the mouth of the bottle and then pulling the scoop down the face of the bottle. As the scoop is dragged down the face of the bottle it will pick up different quantities of solid. Select the smallest sample size at the top and drag the scoop to the beaker on the balance until it snaps in place and then let go. You should have approximately 0.1 g of As_2O_3 in the beaker. Record the mass of the sample in the data table on the following page and return to the laboratory.

3. Drag the beaker from the balance to the sink and hold it under the tap to add a small amount of water. Place it on the stir plate and add the voltmeter probe to the beaker.

4. The buret will be filled with a solution of $KMnO_4$ of unknown concentration. A bottle of permanganate is on the lab bench. Record the unknown sample number. Click the **Save** button in the *Buret Zoom View* window so the titration data can be saved. The horizontal position of the orange handle is off for the stopcock. Open the stopcock by pulling down on the orange handle. The vertical position delivers solution the fastest with three intermediate rates in between. Turn the stopcock to one of the fastest positions. Observe the titration curve. When the blue line begins to turn up, double-click the stopcock to turn it off. Move the stopcock down one position to add volume drop by drop.

   There are two methods for determining the volume at the equivalence point: (1) Stop the titration when a color change occurs. Click the **Stop** button in the *Buret Zoom View*. A blue data link will appear in the lab book. Click the data link icon to open the *Data Viewer* window. Scroll down to the last data entry and record the volume at the equivalence point in the data table. OR (2) Add drops slowly through the equivalence point until the voltages reaches a maximum and levels off. Click the **Stop** button in the *Buret Zoom View*. A data link icon will appear in the lab book. Click the data link icon to open the *Data Viewer* window. Click the **Copy Data** button to copy and paste the data into a spreadsheet program. Plot the first derivative of voltage vs. volume. The peak will indicate the volume at the equivalence point since this is where the voltage is changing the most rapidly as the volume changes.

   To repeat the titration, drag the beaker to the red disposal bucket, move a new beaker from the beaker drawer next to the balance, and repeat the procedure. Do not forget to refill the buret with $KMnO_4$ and place the voltmeter in the beaker each time.

The molecular weight of As_2O_3 is 197.84 g/mol.

Unknown # _____

Data Table

Trial	mass As_2O_3 (g)	volume $KMnO_4$ (mL)	molarity $KMnO_4$ (mole/L)
1			
2			
3			
4			
5			

5. Upon the addition of water, As_2O_3 is converted to H_3AsO_3. During the titration H_3AsO_3 is oxidized to H_3AsO_4 and MnO_4^- is reduced to Mn^{2+}.

   Write a balanced net ionic equation for the reaction.

6. *Using the information in the data table, calculate the molarity of the permanganate solution. (Note that 1 mole of As_2O_3 solid becomes 2 moles of H_3AsO_3 when dissolved in water.)*

7-3: Analysis of a Ferrous Chloride Sample

Titrations provide a method of quantitatively measuring the concentration of an unknown solution. This is done by delivering a titrant of known concentration into an analyte of known volume. (The concentration of an unknown titrant can also be determined by titration with an analyte of known concentration and volume.) In oxidation-reduction (redox) titrations, the voltage resulting from the mixture of an oxidant and reductant can be measured as the titration proceeds. The equivalence point of the titration, or the point where the analyte has been completely consumed by the titrant, is identified by the point where the voltage changes rapidly over a small volume of titrant delivered. In this assignment, you will determine the mass % of an unknown sample of ferrous chloride ($FeCl_2$) by titrating it with a $KMnO_4$ solution of known concentration.

1. Start *Virtual ChemLab,* select *Electrochemistry,* and then select *Analysis of a Ferrous Chloride Sample* from the list of assignments. The lab will open in the Titrations laboratory.

2. You will have to drag the $FeCl_2$ bottle to the front of the lab bench to see the unknown number. Record the $FeCl_2$ Unknown number in the data table on the following page. Click the *Lab Book* to open it. Click the *Beakers* drawer and place a beaker in the spotlight next to the balance. Click on the *Balance* area to zoom in, open the bottle of unknown $FeCl_2$ by clicking on the lid (*Remove Lid*). Drag the beaker to the balance to place it on the balance pan and tare the balance. Pick up the *Scoop* and scoop out some sample by first dragging the scoop to the mouth of the bottle and then pulling the scoop down the face of the bottle. As the scoop is dragged down the face of the bottle it will pick up different quantities of solid. Select the largest sample size and drag the scoop to the beaker on the balance until it snaps in place and then let go. Repeat this again so you have approximately 2 g of unknown in the beaker. Record the unknown number and the mass of the sample in the data table. Return to the laboratory.

3. Place the beaker on the stir plate. Drag the 50 mL graduated cylinder under the tap in the sink and fill it with distilled water. It will automatically snap back into place when it is full. Drag the full 50 mL graduated cylinder to the beaker on the stir plate and then pour the water into the beaker. Now place the voltmeter probe in the beaker and make sure the voltmeter is on.

4. The buret will be filled with 0.0815 M $KMnO_4$. Click the **Save** button in the *Buret Zoom View* window so the titration data can be saved. The horizontal position of the orange handle is off for the stopcock. Open the stopcock by pulling down on the orange handle. The vertical position delivers solution the fastest with three intermediate rates in between. Turn the stopcock to one of the fastest positions. Observe the titration curve. When the blue line begins to turn up, double-click the stopcock to turn it off. Move the stopcock down one position to add volume drop by drop.

    There are two methods for determining the volume at the equivalence point: (1) Stop the titration when a color change occurs. Click the **Stop** button in the *Buret Zoom View*. A blue data link will appear in the lab book. Click the data link icon to open the *Data Viewer* window. Scroll down to the last data entry and record the volume at the equivalence point in the data table. OR (2) Add drops slowly through the equivalence point until the voltage reaches a maximum and levels off. Click the **Stop** button in the *Buret Zoom View*. A data link icon will appear in the lab book. Click the data link icon to open the *Data Viewer* window. Click the **Copy Data** button to copy and paste the data into a spreadsheet program. Plot the first derivative of voltage vs. volume. The peak will indicate the volume at the equivalence point since this is where the voltage is changing the most rapidly as the volume changes.

5. Repeat the titration at least two additional times recording your data in the data table. Do not forget to refill the buret with $KMnO_4$, place the voltmeter probe in the beaker, and add water each time.

    The molecular weight of $FeCl_2$ is 151.91 g/mol.

    Unknown # _____

    Data Table

Trial	mass $FeCl_2$ (g)	volume $KMnO_4$ (mL)	molarity $KMnO_4$ (mol/L)
1			
2			
3			
4			
5			

6. *Write a balanced net ionic equation for the reaction in acidic solution of $FeCl_2$ and $KMnO_4$ (Fe^{2+} is oxidized to Fe^{3+} and MnO_4^- is reduced to Mn^{2+}).*

7. The moles of MnO_4^- can be calculated by multiplying the volume of MnO_4^- required to reach the endpoint multiplied by the molarity of the MnO_4^- solution.

    What are the moles of MnO_4^- used in the titration? _____

8. The moles of $FeCl_2$ can be calculated by using the mole ratio from the balanced equation.

    How many moles of $FeCl_2$ were in the unknown? _____

9. The mass of $FeCl_2$ in the sample can be calculated by multiplying the moles of $FeCl_2$ by the molecular weight of $FeCl_2$.

    What is the mass of $FeCl_2$ in the sample? _____

10. The mass % of $FeCl_2$ in the unknown sample can be calculated by dividing the mass of $FeCl_2$ in the sample by the total mass of the unknown sample.

    What is the % $FeCl_2$ in your unknown sample? _____

11. *What is the average % iron in the unknown sample using your best three answers?*

    _____

8-1: Flame Tests for Metals

Have you ever wondered why a candle flame is yellow? The characteristic yellow of the flame comes from the glow of burning carbon fragments. The carbon fragments are produced by the incomplete combustion reaction of the wick and the candle wax. When elements, such as carbon, are heated to high temperatures, some of their electrons are excited to higher energy levels. When these excited electrons fall back to lower energy levels, they release excess energy in packages of light called photons. The color of the emitted light depends on the individual energy level spacing in the atom. When heated, each element emits a characteristic pattern of photons, which is useful for identifying the element. The characteristic colors of light produced when substances are heated in the flame of a gas burner are the basis for flame tests of several elements. In this assignment, you will perform flame tests that are used to identify several metallic elements.

1. Start *Virtual ChemLab,* select *Descriptive Chemistry,* and then select *Flame Tests for Metals* from the list of assignments. The lab will open in the Inorganic laboratory.

2. On the right side of the stockroom is a shelf labeled *Unknowns*. Click on the *Unknowns* label to create a test tube with an unknown. Now click on each of the following bottles on the shelf: Na^+, K^+, Ca^{2+}, Ba^{2+}, Sr^{2+}, and Cu^{2+}. Do not change the maximum and minimum on the left side and click on the "*Create Unknown*" button. An unknown test tube labeled *Practice* will show up in the blue unknown rack on the unknown shelf. Drag the practice unknown test tube from the blue rack to the blue test tube rack on the lab bench.

3. Drag a test tube from the box and place it on the metal test tube stand. Bottles containing solutions of metal cations are located on the shelves, and you can click on any of the bottles to add them to your test tube. Click on the Na^+ bottle and then drag the test tube to the blue test tube rack. Repeat this process for each new metal cation until you have created test tubes containing Na^+, K^+, Ca^{2+}, Ba^{2+}, Sr^{2+}, and Cu^{2+}.

4. You should note that you have seven test tubes. You will use the two flame test buttons at the bottom of the screen to perform a regular flame test and a flame test with a cobalt filter (blue glass held in front of the flame.) A test tube must be moved from the blue test tube rack to the metal test tube stand in order to perform the flame test. Test tubes dragged from the test tube rack can also be dropped on a test tube in the stand to switch places. You can mouse-over each test tube to identify what metal cation is present in the test tube. As you mouse over each test tube, you will also see a picture of the test tube.

5. Select the test tube containing Na^+ and place it on the metal stand. Click the ***Flame*** button. Record your observations in the data table on the following page. Click the ***Flame w/Cobalt*** button and record your observations in the same table.

6. Drag the K^+ test tube to the metal stand to exchange it with the Na^+. Perform a flame test on K^+ with and without cobalt glass. Record your observations in the data table.

7. For the other four ions, perform a regular flame test only. Do not use the cobalt glass. Record your observations in the data table.

Data Table

Flame Tests	[Answers]
Ion	Flame Color
sodium, Na^+	
sodium, Na^+ (cobalt glass)	
potassium, K^+	
potassium, K^+ (cobalt glass)	
calcium, Ca^{2+}	
barium, Ba^{2+}	
strontium, Sr^{2+}	
copper, Cu^{2+}	
unknown #1	
unknown #2	
unknown #3	
unknown #4	

8. Now perform a flame test on the practice unknown. Determine which of the six metal ions it most closely matches. You may repeat the flame test on any of the six metal ions if necessary. When you are confident that you have identified the unknown, open the *Lab Book* by clicking on it. On the left page, click the **Report** button. On the right page, click on the metal cation that you think is in the practice unknown. Click **Submit**. If all of the cation buttons turn green, you have successfully identified the unknown. If any turn red then you are incorrect. Click on the red disposal bucket to clear all of your samples.

9. Drag the practice unknown to the metal test tube stand. This will randomly create a new unknown. Test and report this unknown. Continue until you have correctly identified four different practice unknowns.

10. The energy of colored light increases in the order red, yellow, green, blue, and violet.

    List the metallic elements used in the flame tests in increasing order of the energy of the light emitted.

11. *What is the purpose of using the cobalt glass in the identification of sodium and potassium?*

    _____

    _____

8-2: Identification of Cations in Solution – Flame Tests

The process of determining the composition of a sample by conducting chemical tests is called qualitative analysis. By using the appropriate tests and applying logic, the identities of the ions present in an unknown solution can be determined. A qualitative analysis scheme is typically made up of a systematic set of chemical reactions where a certain subset of the ions present in the solution are selectively precipitated and removed. The color of the precipitates and solutions provide the means to identify the ions present. Flame tests are also used to identify certain ions that are difficult to identify chemically. In this assignment, you will learn how flame tests are used to identify the cations Na^+ and K^+. As you complete this analysis, remember that careful observation and logical reasoning are the keys to a successful qualitative analysis.

1. Start *Virtual ChemLab*, select *Descriptive Chemistry*, and then select *Identification of Cations in Solution* from the list of assignments. The lab will open in the Inorganic laboratory.

2. Drag a test tube from the box and place it on the metal test tube stand. Now click on the *Na$^+$* bottle to add it to the test and then drag the test tube to the blue test tube rack. Repeat this process with *K$^+$* and a *Na$^+$/K$^+$* mixture. Fill one test tube with just water by clicking on the bottle of distilled water.

3. You should note that you have four test tubes. You can drag your cursor over each test tube to identify what cation is in each test tube and see a picture of what it looks like.

4. In this assignment, you will use the **Flame** and **Flame w/ Cobalt** (cobalt glass held in front of the flame) buttons located at the bottom of the screen. Before performing any chemical test or flame test, a test tube must be moved from the blue test tube rack to the metal test tube stand. If a test tube is already present in the test tube stand, you can drag a test tube from the blue rack to the metal stand to automatically switch places.

5. *Perform a flame test with and without the cobalt filter for just the Na$^+$ solution. Record your*

   observations. _____

6. *Now do the same for just the K$^+$ solution and record your observations.*

   _____

7. *Now do the same for the Na$^+$/K$^+$ mixture and record your observations.* _____

   _____

8. *Now do the same with just water to get a feel for what it looks like with no chemicals other than*

   water. Record your observations. _____

9. Click on the *Unknowns* label to reconfigure the laboratory to create a practice unknown. Click on the *Na$^+$* and *K$^+$* bottles and on the holographic display on the left make the minimum = 0 and maximum = 2. Click on the *"Create Unknown"* button. An unknown test tube labeled *Practice* will be placed in the blue unknown rack on the stockroom counter. Drag the practice unknown from the blue rack to the blue test tube rack on the lab bench.

10. Perform a flame test with and without the cobalt filter and determine if the practice unknown contains sodium or potassium or both or neither. To check your results, click on the *Lab Book* and on the left page, click the ***Report*** button, select the ions you think are present in the unknown, click ***Submit***. If the ion button is green, you correctly determined whether the ion was present or not. If the ion button is red you did not make the correct analysis. Return to the laboratory and click the red disposal bucket to clear the lab. If you want to repeat with a new practice unknown, return to the stockroom and retrieve it from the blue rack. Continue until you obtain only green buttons when submitted.

11. After you have finished with your practice unknowns, you are now ready to accept your assigned unknown. Clear the laboratory of any test tubes by clicking on the red disposal bucket. Click on the clipboard on the counter. The clipboard contains a list of test tubes containing known and unknown solutions. The possible cations in each unknown are shown in the summary when mousing over the title. Select *Basic Knowns and Unknowns* and then *Alkali Metals Unknown*, which will give you an unknown that could contain any combination of Na^+ and K^+ or both or neither. Test your assigned unknown and report the cations found in your unknown in the space below. Don't forget to record your unknown number.

Unknown _____ Cations found: _____

8-3: Identification of Cations in Solution — Ag^+, Hg_2^{2+}, Pb^{2+}

The process of determining the composition of a sample by conducting chemical tests is called qualitative analysis. By using the appropriate tests and applying logic, the identities of the ions present in an unknown solution can be determined. A qualitative analysis scheme is typically made up of a systematic set of chemical reactions where a certain subset of the ions present in the solution are selectively precipitated and removed. The color of the precipitates and solutions provide the means to identify the ions present. Flame tests are also used to identify certain ions that are difficult to identify chemically. In this assignment, you will learn the basics of a qualitative analysis scheme by performing an analysis on a mixture of Ag^+, Hg_2^{2+}, and Pb^{2+}. As you complete this analysis, remember that careful observation and logical reasoning are the keys to a successful qualitative analysis.

1. Start *Virtual ChemLab,* select *Descriptive Chemistry,* and then select *Identification of Cations in Solution* from the list of assignments. The lab will open in the Inorganic laboratory.

2. Drag a test tube from the box and place it on the metal test tube stand. Now click on the Ag^+, Hg_2^{2+}, and Pb^{2+} bottles to add these cations to the test tube and then drag the test tube to the blue test tube rack. (There is Hg^{2+} and Hg_2^{2+} on the shelf. Make sure you obtain Hg_2^{2+}.) As you proceed with the chemical analysis watch the screen on the left to see the chemistry involved in the chemical reactions. You may also want to make some copies of your original test tube by clicking on the ***Divide*** button in case you make a mistake and need to start over.

3. Move the test tube to the metal stand. Click the NaCl reagent bottle to add chloride to the test tube.

    What observations can you make? _____

    *Click the **Centrifuge** button. What observations can you make?* _____

    Each of the three ions form insoluble precipitates (solids) with chloride. If the solution turns cloudy white it indicates that at least one of the three ions is present. Now, we must determine which one. (Remember we already know what is in the test tube, but for an unknown you won't.)

4. Turn the heat on with the ***Heat*** button. Observe the screen, on the left.

    What happened? If you cannot tell, turn the heat on and off while observing the TV screen.

    _____

    With the heat turned on, click ***Centrifuge*** again and then ***Decant***. (Decanting separates the solids from the solution and places the solution in the test tube rack.) Drag your cursor over the new test tube in the rack.

    What appears on the screen? What appears in the picture window? _____

    _____

    This is the test for Pb^{2+}. If heated, it is soluble. When cooled it becomes insoluble.

5. Turn off the ***Heat***. With the test tube containing the two remaining ions in the metal stand, click the NH_3 bottle on the reagent shelf.

What do you observe? _____

Addition of ammonia produces a diammine silver complex ion that is soluble. The mercury produces a green/black solid. This is the test for mercury.

6. ***Centrifuge*** and then ***Decant*** to pour the silver ion into another test tube. Move the tube with the black mercury solid to the red disposal bucket. Move the test tube containing silver back to the metal stand. Click the pH 4 reagent bottle to make the solution slightly acidic.

   What do you observe? _____

   _____

   The silver ion is soluble as the diammine silver complex ion in pH 10 and is insoluble as AgCl in pH 4. You can click alternately on each of the pH bottles to confirm this test for the silver ion.

7. Click on the *Unknowns* label to reconfigure the laboratory to create a practice unknown. Click on the Ag^+, Hg_2^{2+}, and Pb^{2+} bottles and on the holographic display on the left make the minimum = 0 and maximum = 3. Click on the *"Create Unknown"* button. An unknown test tube labeled *Practice* will be placed in the blue unknown rack on the stockroom counter. Drag the practice unknown from the blue rack to the blue test tube rack on the lab bench.

8. Test the *Practice* unknown and determine if it contains each of the ions Ag^+, Hg_2^{2+}, and Pb^{2+}. To check your results, click on the *Lab Book* and on the left page, click the ***Report*** button, select the ions you think are present in the unknown, click ***Submit***. If the ion button is green, you correctly determined whether the ion was present or not. If the ion button is red you did not make the correct analysis. Return to the laboratory and click the red disposal bucket to clear the lab. If you want to repeat with a new practice unknown, return to the stockroom and retrieve it from the blue rack. Continue until you obtain only green buttons when submitted.

9. After you have finished with your practice unknowns, you are now ready to accept your assigned unknown. Clear the laboratory of any test tubes by clicking on the red disposal bucket. Click on the clipboard on the counter. The clipboard contains a list of test tubes containing known and unknown solutions. The possible cations in each unknown are shown in the summary when mousing over the title. Select *Basic Knowns and Unknowns* and then *The Chlorides Unknown*, which will give you an unknown that could contain any combination of Ag^+, Hg_2^{2+}, and Pb^{2+} or all of them or just water. Test your assigned unknown and report the cations found in your unknown in the space below. Don't forget to record your unknown number.

Unknown _____   Cations found: _____

8-4: Identification of Cations in Solution – Co^{2+}, Cr^{3+}, Cu^{2+}

The process of determining the composition of a sample by conducting chemical tests is called qualitative analysis. By using the appropriate tests and applying logic, the identities of the ions present in an unknown solution can be determined. A qualitative analysis scheme is typically made up of a systematic set of chemical reactions where a certain subset of the ions present in the solution are selectively precipitated and removed. The color of the precipitates and solutions provide the means to identify the ions present. Flame tests are also used to identify certain ions that are difficult to identify chemically. In this assignment, you will learn the basics of a qualitative analysis scheme by performing an analysis on a mixture of Co^{2+}, Cr^{3+}, and Cu^{2+}. As you complete this analysis, remember that careful observation and logical reasoning are the keys to a successful qualitative analysis.

1. Start *Virtual ChemLab*, select *Descriptive Chemistry,* and then select *Identification of Cations in Solution* from the list of assignments. The lab will open in the Inorganic laboratory.

2. Drag a test tube from the box and place it on the metal test tube stand. Now click on the *Co^{2+}, Cr^{3+},* and *Cu^{2+}* bottles to add these cations to the test tube and then drag the test tube to the blue test tube rack. As you proceed with the chemical analysis watch the screen, on the left to see the chemistry involved in the chemical reactions. You may also want to make some copies of your original test tube by clicking on the **Divide** button in case you make a mistake and need to start over.

3. *Move the test tube to the metal stand. Click the NaOH bottle on the reagent shelf. What observations can you make?* _____

    _____

4. Click **Centrifuge** and **Decant**. (Decanting separates the solids from the solution and places the solution in the test tube rack.)

    What observations can you make as you drag your cursor over each test tube? _____

    _____

    This is the test for chromium. If the new test tube in the blue rack is green when decanted then chromium is present. You can confirm it by placing the clear green test tube in the metal stand and clicking pH 10 and then adding HNO$_3$.

    What observations can you make? _____

    _____

5. With the test tube containing the cobalt and copper precipitate in the metal stand, add NH$_3$.

    What observations can you make? _____

    _____

6. **Centrifuge** and **Decant**. Add HNO$_3$ to the tube in the metal stand containing the precipitate.

What observations can you make? _____

_____

This is the confirmatory test for cobalt ion (Co^{2+}).

7. Place the test tube from the blue rack, which is the solution from step # 5, in the metal stand. Add HNO_3.

    What observations can you make? _____

    _____

    This is the confirmatory test for copper.

8. Click on the *Unknowns* label to reconfigure the laboratory to create a practice unknown. Click on the Co^{2+}, Cr^{3+}, and Cu^{2+} bottles and on the holographic display on the left make the minimum = 0 and maximum = 3. Click on the *"Create Unknown"* button. An unknown test tube labeled *Practice* will be placed in the blue unknown rack on the stockroom counter. Drag the practice unknown from the blue rack to the blue test tube rack on the lab bench.

9. Test the *Practice* unknown and determine if it contains each of the ions Co^{2+}, Cr^{3+}, and Cu^{2+}. To check your results, click on the *Lab Book* and on the left page, click the **Report** button, select the ions you think are present in the unknown, click **Submit**. If the ion button is green, you correctly determined whether the ion was present or not. If the ion button is red you did not make the correct analysis. Return to the laboratory and click the red disposal bucket to clear the lab. If you want to repeat with a new practice unknown, return to the stockroom and retrieve it from the blue rack. Continue until you obtain only green buttons when submitted.

10. After you have finished with your practice unknowns, you are now ready to accept your assigned unknown. Clear the laboratory of any test tubes by clicking on the red disposal bucket. Click on the clipboard on the counter. The clipboard contains a list of test tubes containing known and unknown solutions. The possible cations in each unknown are shown in the summary when mousing over the title. Select *Basic Knowns and Unknowns* and then *Transition Metals (I) Unknown*, which will give you an unknown that could contain any combination of Co^{2+}, Cr^{3+}, and Cu^{2+} or all of them or just water. Test your assigned unknown and report the cations found in your unknown in the space below. Don't forget to record your unknown number.

    Unknown _____   Cations found: _____

8-5: Identification of Cations in Solution – Ba^{2+}, Sr^{2+}, Ca^{2+}, Mg^{2+}

The process of determining the composition of a sample by conducting chemical tests is called qualitative analysis. By using the appropriate tests and applying logic, the identities of the ions present in an unknown solution can be determined. A qualitative analysis scheme is typically made up of a systematic set of chemical reactions where a certain subset of the ions present in the solution are selectively precipitated and removed. The color of the precipitates and solutions provide the means to identify the ions present. Flame tests are also used to identify certain ions that are difficult to identify chemically. In this assignment, you will need to develop your own qualitative analysis scheme to separate and identify the Group II cations Ba^{2+}, Sr^{2+}, Ca^{2+}, and Mg^{2+}. As you complete this analysis, remember that careful observation and logical reasoning are the keys to a successful qualitative analysis.

1. Start *Virtual ChemLab,* select *Descriptive Chemistry,* and then select *Identification of Cations in Solution* from the list of assignments. The lab will open in the Inorganic laboratory.

2. Drag a test tube from the box and place it on the metal test tube stand. Now click on the Ba^{2+}, Sr^{2+}, Ca^{2+}, and Mg^{2+} bottles to add these cations to the test tube and then drag the test tube to the blue test tube rack. As you proceed with the chemical analysis watch the screen on the left to see the chemistry involved in the chemical reactions. You may also want to make some copies of your original test tube by clicking on the **Divide** button in case you make a mistake and need to start over.

3. *What do the solubility rules tell you about the way to separate the second group of cations? How can you tell each of the ions in this group apart? Think about changing the temperature and the pH. Design your own qualitative scheme for identification of these four Group II ions and write it below. Experiment with each of the four ions alone and then in combinations. Test an unknown to see if you can really determine the presence or absence of each of the four ions in this group.*

4. Click on the *Unknowns* label to reconfigure the laboratory to create a practice unknown. Click on the Ba^{2+}, Sr^{2+}, Ca^{2+}, and Mg^{2+} bottles and on the holographic display on the left make the minimum = 0 and maximum = 4. Click on the *"Create Unknown"* button. An unknown test tube labeled *Practice* will be placed in the blue unknown rack on the stockroom counter. Drag the practice unknown from the blue rack to the blue test tube rack on the lab bench.

5. Test the *Practice* unknown and determine if it contains each of the ions Ba^{2+}, Sr^{2+}, Ca^{2+}, and Mg^{2+}. To check your results, click on the *Lab Book* and on the left page, click the **Report** button, select the ions you think are present in the unknown, click **Submit**. If the ion button is green, you correctly determined whether the ion was present or not. If the ion button is red you did not make the correct analysis. Return to the laboratory and click the red disposal bucket to clear the lab. If you want to repeat with a new practice unknown, return to the stockroom and retrieve it from the blue rack. Continue until you obtain only green buttons when submitted.

6. After you have finished with your practice unknowns, you are now ready to accept your assigned unknown. Clear the laboratory of any test tubes by clicking on the red disposal bucket. Click on the clipboard on the counter. The clipboard contains a list of test tubes containing known and unknown solutions. The possible cations in each unknown are shown in the summary when mousing over the title. Select *Basic Knowns and Unknowns* and then *Alkaline Earth Metals Unknown*, which will give you an unknown that could contain any combination of Ba^{2+}, Sr^{2+}, Ca^{2+}, and Mg^{2+} or all of them or just water. Test your assigned unknown and report the cations found in your unknown in the space below. Don't forget to record your unknown number.

Unknown _____   Cations found: _____

8-6: Identification of Cations in Solution – Co^{2+}, Cu^{2+}, Ni^{2+}

The process of determining the composition of a sample by conducting chemical tests is called qualitative analysis. By using the appropriate tests and applying logic, the identities of the ions present in an unknown solution can be determined. A qualitative analysis scheme is typically made up of a systematic set of chemical reactions where a certain subset of the ions present in the solution are selectively precipitated and removed. The color of the precipitates and solutions provide the means to identify the ions present. Flame tests are also used to identify certain ions that are difficult to identify chemically. In this assignment, you will need to develop your own qualitative analysis scheme to separate and identify the transition metal cations Co^{2+}, Cu^{2+}, and Ni^{2+}. As you complete this analysis, remember that careful observation and logical reasoning are the keys to a successful qualitative analysis.

1. Start *Virtual ChemLab,* select *Descriptive Chemistry,* and then select *Identification of Cations in Solution* from the list of assignments. The lab will open in the Inorganic laboratory.

2. Drag a test tube from the box and place it on the metal test tube stand. Now click on the *Co²⁺, Cu²⁺,* and *Ni²⁺* bottles to add these cations to the test tube and then drag the test tube to the blue test tube rack. As you proceed with the chemical analysis watch the screen on the left to see the chemistry involved in the chemical reactions. You may also want to make some copies of your original test tube by clicking on the **Divide** button in case you make a mistake and need to start over.

3. *What do the solubility rules tell you about the way to separate the second group of cations? How can you tell each of the ions in this group apart? Think about changing the temperature and the pH. Design your own qualitative scheme for identification of these three transition metal ions and write it below. Experiment with each of the three ions alone and then in combinations. Test an unknown to see if you can really determine the presence or absence of each of the three ions in this group.*

4. Click on the *Unknowns* label to reconfigure the laboratory to create a practice unknown. Click on the *Co²⁺, Cu²⁺,* and *Ni²⁺* bottles and on the holographic display on the left make the minimum = 0 and maximum = 3. Click on the *"Create Unknown"* button. An unknown test tube labeled *Practice* will be placed in the blue unknown rack on the stockroom counter. Drag the practice unknown from the blue rack to the blue test tube rack on the lab bench.

5. Test the *Practice* unknown and determine if it contains each of the ions Co^{2+}, Cu^{2+}, and Ni^{2+}. To check your results, click on the *Lab Book* and on the left page, click the **Report** button, select the ions you think are present in the unknown, click **Submit**. If the ion button is green, you correctly determined whether the ion was present or not. If the ion button is red you did not make the correct analysis. Return to the laboratory and click the red disposal bucket to clear the lab. If you want to repeat with a new practice unknown, return to the stockroom and retrieve it from the blue rack. Continue until you obtain only green buttons when submitted.

6. After you have finished with your practice unknowns, you are now ready to accept your assigned unknown. Clear the laboratory of any test tubes by clicking on the red disposal bucket. Click on the clipboard on the counter. The clipboard contains a list of test tubes containing known and unknown solutions. The possible cations in each unknown are shown in the summary when mousing over the title. Select *Basic Knowns and Unknowns* and then *Transition Metals (II) Unknown*, which will give you an unknown that could contain any combination of Co^{2+}, Cu^{2+}, and Ni^{2+} or all of them or just water. Test your assigned unknown and report the cations found in your unknown in the space below. Don't forget to record your unknown number.

Unknown _____   Cations found: _____

Additional Assignments

The following worksheets contain additional assignments that are more advanced, explore topics with more depth, or introduce you to the more detailed parts of the *Virtual ChemLab* simulations. Keep in mind that these assignments are *not* contained on the whiteboard found in the virtual laboratory. To perform these assignments, you must enter the general chemistry laboratory and then select the appropriate laboratory workbench as indicated in the assignment. In general, there are fewer step-by-step directions in these assignments and it is assumed you are familiar with the laboratory interface.

Inert Salts

Na$^+$ and Cl$^-$ are often referred to as spectator ions. This is because during a simple reaction like a titration of sodium hydroxide with hydrochloric acid, they do not affect the equivalence point or pH of the reaction. These ions are stable enough that they are not affected by the pH. They exist as separate salts. Although inert salts do not affect the pH of titrations, they do affect the conductivity of solutions. In this assignment you will look at how inert salts affect the conductivity of acid/base titrations.

1. Take out Preset Experiment 1. Run the titration and graph the pH and conductivity curves on the following graph. Label the axis.

2. *Describe the shape of the conductivity curve. Why is it shaped the way it is?* _____

_____

_____

_____

3. Take out Preset Experiment 1 again. Perform the same titration after adding approximately 3.0 g NaCl. In a different color draw the new pH and conductivity curves on the graph above.

4. *How did the conductivity of the titration change with the addition of NaCl? Why?*

_____

_____

_____

Graphing Titration Data

There are many ways to determine the equivalence point of a titration. The most common way is to use the end point of the titration, found by using an indicator or noticing a color change in a redox titration, as an estimate for the equivalence point. If an appropriate indicator is used, its accuracy can be close. However, a more accurate way to determine the equivalence point is to use pH versus volume data to find the point at which the slope of the line of the titration is greatest. In this assignment, you will use a spreadsheet to accurately calculate the equivalence point of a titration.

Take out Preset Experiment 5. The calibrated pH meter is in the beaker, the graph window is open, and 0.9062 g of the $NaHCO_3$ solid is in the beaker. The conductivity meter is also turned on and in the beaker, but will not be used for this assignment. The indicator Methyl orange is in the analyte beaker. Calculate where you expect the equivalence point to be. Open your lab book and save the titration information as it nears and passes the expected equivalence point. With the titration of a weak base, it is best to save the data when the stopcock is on the slowest or second to slowest setting. Open the link in the lab book and copy the data to a spreadsheet. Create a graph showing the pH as a function of volume. It should look similar to the graph in the graph window.

In another column in your spreadsheet, calculate the slope between the individual titration points. This can be done by dividing the difference between the current and previous pHs by the difference between the current and previous volumes. Graph the line of these calculations on the same graph as the graph of the pH as a function of volume. The point at which this line reaches the highest point is the equivalence point of the titration. Include this graph with this assignment.

1. *Why does this method work to accurately determine the equivalence point?* _____

   _____

2. *What is the equivalence point of the titration?* _____

3. *What is the percentage difference between the equivalence point you calculated and the expected*

   *equivalence point from your preliminary calculation?*_____

4. *Why is this method more accurate than using an indicator to estimate the equivalence points?*

   _____

   _____

5. *Is this technique more useful with titrations of a strong acid and base or titrations where either the*

   acid or the base is weak? Why? _____

   _____

   _____

Activities

"Activity" is a term that helps us understand how a substance will react in circumstances different than that of a selected reference state. Activities are necessary for precise calculations of equilibrium solutions because the ionic strength of dissociated molecules has some effect on the equilibrium. In order to account for activities, a more accurate form of the equilibrium constant equation is needed. This equation is $K = \dfrac{[C]^c \gamma_C^c [D]^d \gamma_D^d}{[A]^a \gamma_A^a [B]^b \gamma_B^b}$. The activity coefficient γ equals one in the reference state, which is the ideal state. As the solutions deviates from the reference state, the activity coefficients also deviate from one. Thus, γ measures the deviation from the ideal state. You will not be using this equation to make calculations, but note that in the stockroom of the titration laboratory, there is a switch where you can decide whether or not activities affect your solution. In this assignment you will learn more about activities and in what circumstances they most affect a solution.

1. Take out Preset Experiment 1. This is a strong acid strong base titration. There are 25 mL of HCl in the beaker. Perform the titration with the activity coefficients on and off. Use data from these titrations to calculate the concentration of NaOH as though it were an unknown.

   *How much of a difference do the activities make in these calculations?*_____

2. Design a titration using KHP and NaOH. Since KHP is a solid, you can determine the molarity of the solution when you mix it with water. Mix a solution where the concentration of the solution is 3.0 M or greater. Titrate this solution with NaOH when the activity coefficients are off. Calculate the concentration of NaOH as if it were an unknown. Repeat this procedure with the activity coefficients on. Show your calculations for the concentration of NaOH.

3. *What difference do the activities make when the KHP is in a more concentrated solution?*

_____

Indicators

Indicators are chemical dyes that change color over a specific range of pH. For example, litmus changes from red to blue as the acid is converted to its conjugate base. The best indicators have intense colors so that only a few drops of dilute indicator will cause the color to change. Indicators provide a way to estimate the equivalence point of a titration, which is when the moles of analyte have been consumed by moles of titrant. The point at which the indicator changes color is called the end point. In this assignment, you will learn more about using indicators.

1. Take out Preset Experiment 1. This is a strong acid strong base titration. *Look at the indicator chart and list three indicators that would be appropriate for this titration.* If you are not sure where the equivalence point is, perform the titration with the pH meter.

    _____

    _____

2. *Graph a titration curve for this titration and show where each of the three indicators you chose change colors on the following graph.*

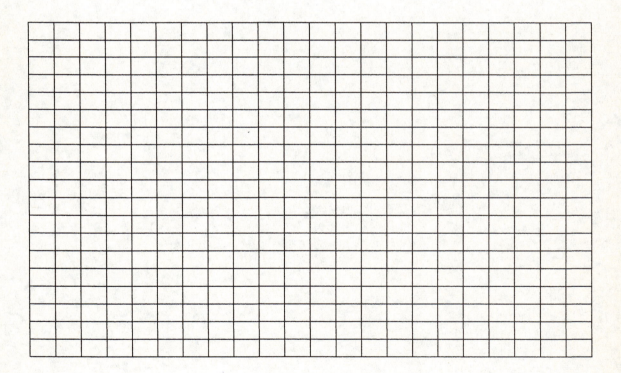

3. Now take out Preset Experiment 3. This is a weak acid strong base titration. *Which indicators would*

    be appropriate for this titration? _____

4. The end points of indicators are only estimates of the equivalence point. *What characteristics of an indicator need to be considered to choose one that will give the most accurate estimate possible?*

_____

_____

Will indicator end points generally be more accurate with a strong acid strong base titration or one

in which the acid or base is weak? Why? _____

_____

_____

_____

Buoyancy

In a titration lab, there are many factors that can affect the accuracy of your measurements. Since small errors can make a difference in the results, it is important to eliminate all sources of error that you have control over. In this assignment, you will learn how to compensate for a buoyancy error that occurs when a substance is weighed on the balance.

The buoyancy equation is $m = m_{obs} \dfrac{\left(1 - \dfrac{d_a}{d_w}\right)}{\left(1 - \dfrac{d_a}{d}\right)}$ where m_{obs} is the observed mass from the balance, d_a is the density of air, d_w is the density of the standard weights, and d is the density of the substance being weighed. m is the corrected (actual) mass of the sample.

1. A buoyancy error occurs when the density of the substance being weighed is different than the density of the standard weights that were used to calibrate the balance. *Why is error introduced into the measurement if the density of the substance being weighed and the density of the substance used to calibrate the balance are not the same?* _____

    _____

    _____

2. Take out KHP from the stockroom. Weigh out approximately 2.00 g of KHP. *Record the mass of KHP you measured out according to the balance.* _____

3. The density of air is 0.0012 g/mL near 760 torr and 25° C and the density of the weights is 8.0 g/mL. The density of KHP is 1.64 g/mL. *Using the buoyancy equation, what is the corrected mass of the sample you measured out?*

    _____

4. *What is the percentage difference in moles of KHP between the measured and corrected mass of the sample?* The molecular weight of KHP is 204.22 g/mol. _____

5. *How will the calculation of an equivalence point in an acid base titration be affected if buoyancy corrections are not made?* _____

Glassware Calibration

The glassware you will use in the virtual chemistry laboratory is as individual to you as your own glassware in a real laboratory. There is an error in your glassware similar to the glassware error in a real laboratory. To have the most accurate calculations in your work, you must calibrate your pipet and buret to receive your tolerance level of your glassware. The error you calculate will be applicable each time you use that glassware.

The density of water at 25°C is 0.9970479g/ml. With this information, you can calculate the exact volume of water delivered if you know the mass of water delivered.

1. Open the buret window and lab book. Fill the buret with water. Tare a beaker by placing it on the balance and clicking on *Tare*. Place the tared beaker under the buret and deliver 10 mL of water into the beaker. Record the exact volume of water delivered below according to the buret reading to the nearest 0.02 mL. *Take the beaker to the balance and record the mass of water delivered. Do this four times, taring the beaker before each new addition of water from the buret.* When you have finished, you should have delivered approximately 40 mL from the buret.

    Volume #1: _____   Mass #1: _____

    Volume #2: _____   Mass #2: _____

    Volume #3: _____   Mass #3: _____

    Volume #4: _____   Mass #4: _____

2. *Calculate the actual volume delivered each time using the density of water.*

    Volume #1: _____   Volume #3: _____

    Volume #2: _____   Volume #4: _____

3. *Calculate the correction in mL at each volume delivered.* (The correction is the amount you have to add or subtract from the amount delivered according buret reading in order to know the correct volume delivered.)

    Volume #1: _____   Volume #3: _____

    Volume #2: _____   Volume #4: _____

4. *Plot the points for the total volume delivered from the buret versus on the x-axis versus the correction in mL at each volume on the y-axis. Connect the points to show the correction at each 10 mL interval.*

All glassware needs to be calibrated in order to be used accurately in the lab. Now you are going to calibrate the 25 mL pipet.

5. Open the pipet drawer. Double click on the 25 ml pipet. Take a beaker out and add some water. Fill the pipet and move the beaker of water. Tare a different beaker by placing it on the balance and clicking on *Tare*. Place the tared beaker under the pipet to empty the pipet into the beaker. *Take the beaker to the balance and record the mass of water. Do this three times.*

   Mass #1: _____   Mass #2: _____   Mass #3: _____

6. *Calculate the actual volume delivered each time.*

   Volume #1: _____   Volume #2: _____   Volume #3: _____

7. *What is the average correction required for your 25 mL pipet?* _____

Boyle's Law: 1/Volume versus Pressure – 1

Name_____

Robert Boyle, a philosopher and theologian, studied the properties of gases in the seventeenth century. He noticed that gases behave similarly to springs; when compressed or expanded, they tend to 'spring' back to their original volume. He published his findings in 1662 in a monograph entitled "The Spring of the Air and Its Effects." You will make observations similar to those of Robert Boyle.

The purpose of this experiment is to learn more about the relationship between the pressure and volume of an ideal gas. You will do this by graphing values of $1/V$ versus P. Graphing the inverse of the volume will make the relationship between P and V easier to see graphically. In order to obtain data to graph, you will be changing pressure as all other variables except for volume are kept constant.

Take out Preset Experiment 1. Open your lab book and save the data as you increase the pressure from 1 atm. Calculate $1/V$ for each of the corresponding volumes at each pressure. These are the values you will use to make your graph. Graph $1/V$ on the y-axis versus P on the x-axis on the chart below. Repeat the procedure with N_2 (Preset Experiment 2) and with a van der Waals gas with the a and b parameters set to N_2 (Preset Experiment 8). Your graph should at least cover the maximum pressure of N_2. Label the graph.

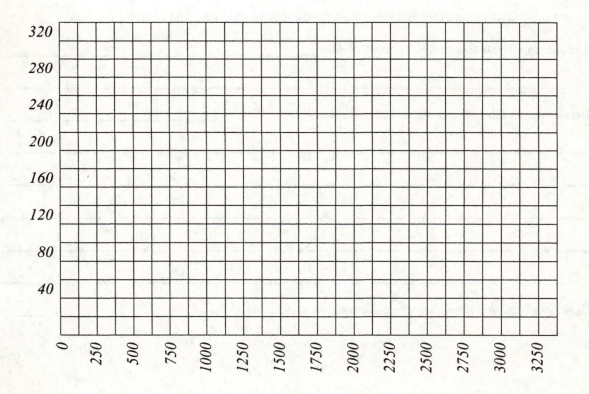

P (atm)

1. *Which line is straight?* _____

2. *What is the relationship between P and V? Use the straight line to explain.* _____

   _____

   _____

   What affects the slope of this line? _____

3. *Name some of the reasons that the other lines are not straight.* _____

   _____

   _____

   _____

   _____

   _____

4. The van der Waals parameters that were used for these data points are the ones for N_2. *Where*

   is the van der Waals line closest to the N_2 line? _____

   _____

5. *Where is the van der Waals line not close to the N_2 line? Why?* _____

   _____

   _____

   _____

   _____

6. *Where is the van der Waals approximation the most accurate?* _____

Boyle's Law: 1/Volume versus Pressure – 2

Robert Boyle, a philosopher and theologian, studied the properties of gases in the seventeenth century. He noticed that gases behave similarly to springs; when compressed or expanded, they tend to 'spring' back to their original volume. He published his findings in 1662 in a monograph entitled "The Spring of the Air and Its Effects." You will make observations similar to those of Robert Boyle.

The purpose of this experiment is to learn more about the relationship between the pressure and volume of an ideal gas. You will do this by graphing values of $1/V$ versus P. Graphing the inverse of the volume will make the relationship between P and V easier to see in the form of a graph. In order to obtain data to graph, you will be changing pressure as all other variables except for volume are kept constant.

Choose an experiment where pressure or volume is dependent on the other and then save the data in your lab book as you increase or decrease the independent variable. Use this data to calculate $1/V$ for each data point. Complete this procedure for an ideal gas, CO_2, He, and N_2. Graph $1/V$ on the y-axis versus P on the x-axis for each gas on a spreadsheet. Include in your graph the maximum pressures of all of the real gases. Make sure that all parameters of the gases are the same when you save the data so the lines can be compared. Label your graph. *(An example of the required graph is included at the end of the assignment.)*

1. *Which line is straight?* _____

2. *What affects the slope of this line?* _____

3. *Why aren't the other lines straight?* _____

   _____

4. *Where are the lines most similar to the straight line?* _____

   _____

5. *Where are they different?* _____

   _____

6. *From your graph, what can you say in general about real gases as the pressure increases?*

   _____

7. *Use your graph to describe the behavior of N_2 and helium.* Note that the *a* parameter of the van der Waals gas for Helium is 0.0341 and that of N_2 is 1.39 $L^2 \cdot atm \cdot mol^{-2}$.

   _____

   _____

   _____

8. *Why does the line for CO_2 slope above the ideal line and the lines for N_2 and Helium slope*

   below the ideal line? _____

   _____

9. *Explain why some real gases behave more like an ideal gas than other real gases.*

   _____

   _____

   _____

   _____

10. *What is another way to graph values derived from P and V so that the graph will be a*

    straight line? _____

Compressibility

The compressibility of a gas, $z=PV/nRT$, is an effective method for comparing the behavior of real gases to that of an ideal gas. For an ideal gas z is always equal to one, therefore, deviations from one are a measure of the non-ideality of the gas. You will use this compressibility factor to make observations about the behavior of real gases.

Take out the experiment where volume is dependent on pressure and temperature. For this assignment you will be using N_2. Set the temperature to 150 K and adjust the pressures to those listed in the table below. For each pressure, calculate the compressibility and list it in the table. Calculate the compressibilities for the pressures listed at 250 K and 1000 K. *Using the data from the table, graph compressibility on the y-axis versus pressure for the data of each temperature on the graph below. Label your graph.*

150 K

P (atm)	z
20	
80	
500	
1000	
1200	

250 K

P (atm)	z
20	
200	
500	
1000	
1200	

1000 K

P (atm)	z
20	
200	
500	
1000	
1200	

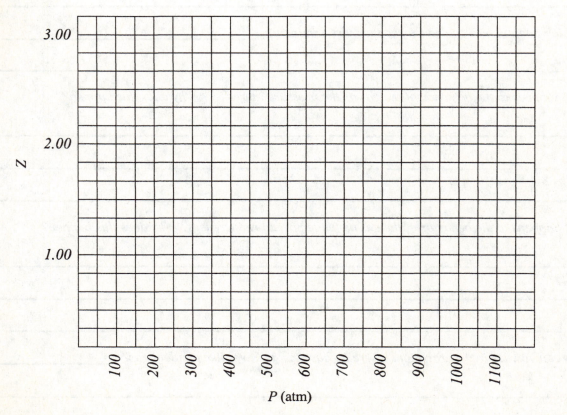

1. *What causes the ratio PV/nRT to go below one for a nonideal gas?* _____

_____

_____

2. *What causes the ratio PV/nRT to go above one for a nonideal gas?* _____

_____

_____

3. *What is happening when the line's slope changes from negative to positive? Why does this happen?*

_____

_____

_____

_____

_____

4. *At which point is the repulsive force the most dominant?* _____

_____

5. *What happens to the behavior of the gas as the pressure increases? Why does this happen?*

_____

_____

_____

6. *What happens to the behavior of the gas as the temperature increases? Why does this happen?*

_____

_____

_____

7. *What can you say in general about where N_2 gas most behaves like an ideal gas?* _____

_____

8. *How would the compressibility be affected if you used m^3 instead of L for volume? What if you used degrees Celsius for temperature?* _____

_____

_____

Van der Waals Gases – 1

The van der Waals gas equation is a closer approximation to the behavior of real gases than the ideal gas equation because it accounts for attractive and repulsive forces. The equation for a van der Waals gas is $\left(P + a\dfrac{n^2}{V^2}\right)(V - nb) = nRT$. The term that includes the parameter a accounts for the attractive

interactions of the gas and the b term accounts for repulsive interactions. These parameters are unique for every gas and are determined by fitting experimental PVT data of real gases. The purpose of this experiment is to learn more about the van der Waals equation and its relationship to real gases. You will do this by rearranging the van der Waals equation and then applying it to real gas conditions in order to observe specific properties of van der Waals gases as compared to real gases.

1. The compressibility of a gas, $z=PV/nRT$, is an effective method for comparing the behavior of real gases to that of an ideal gas. For an ideal gas z is always equal to one, therefore, deviations from one are a measure of the non-ideality of the gas. Solve for the compressibility of the van der Waals equation and then use the equation to answer the following two questions. It may be easier to first solve for the pressure of the van der Waals gas.

2. *Explain how the attractive interactions of a van der Waals gas affect its compressibility.*

    _____

3. *Explain how the repulsive forces of a van der Waals gas affect its compressibility.*

    _____

    _____

Check how close van der Waals gases are to their real gas counterparts. Take out the experiment where pressure is dependent on volume and temperature. You will be using N_2 gas and the N_2 van der Waals gas with parameters $a = 1.390$ atm· L^2·mol^{-2} and $b = 0.03913$ L·mol^{-1}. Set the variables to 1.000 mol and 100 K and calculate the compressibility at the volumes listed in the tables below.

Compare this to the compressibility of the van der Waals gas and calculate the percentage difference between the two values.

Volume (L)	Z N_2	Z vdW	% diff.
4.000			
3.000			
2.000			

Calculate the following values at 500 K.

Volume (L)	Z N_2	Z vdW	% diff.
4.000			
2.000			
1.000			

Calculate the following values at 1200 K.

Volume (L)	Z N_2	Z vdW	% diff.
4.000			
2.000			
1.000			

4. *What is the general trend of the difference between the compressibility of the real gas and the compressibility of the van der Waal gas as the temperature increases? Why?*

_____

_____

_____

5. *What is the general trend as the volume decreases? Why?* _____

_____

_____

_____

6. *Under what conditions is the van der Waals equation the best approximation for the behavior of the real gas N_2?* _____

Van der Waals Gases – 2

The van der Waals gas equation is a closer approximation to the behavior of real gases than the ideal gas equation because it accounts for attractive and repulsive forces. The equation for a van der Waals gas

is $\left(P + a\dfrac{n^2}{V^2} \right)(V - nb) = nRT$. The term that includes the parameter a accounts for the attractive

interactions of the gas and the b term accounts for repulsive interactions. These parameters are unique for every gas and are determined by fitting experimental PVT data of real gases. The purpose of this experiment is to learn more about the van der Waals equation and its relationship to real gases. You will do this by rearranging the van der Waals equation and then applying it to real gas conditions in order to observe specific properties of van der Waals gases as compared to real gases.

1. The compressibility of a gas, $z=PV/nRT$, is an effective method for comparing the behavior of real gases to that of an ideal gas. For an ideal gas z is always equal to one, therefore, deviations from one are a measure of the non-ideality of the gas. Solve for the compressibility of the van der Waals equation and then use the equation to answer the following two questions. It may be easier to first solve for the pressure of the van der Waals gas.

2. *Explain how the attractive interactions of a van der Waals gas affect its compressibility.*

   _____

3. *Explain how the repulsive interactions of a van der Waals gas affect its compressibility.*

   _____

   _____

Check how close van der Waals gases are to their real gas counterparts. Take out the experiment where pressure is dependent on volume and temperature. You will be using N_2 gas and the N_2 van der Waals gas with parameters $a = 1.390$ atm· L^2·mol^{-2} and $b = 0.03913$ L·mol^{-1}. Set the variables to 1.000 mol and 100 K and calculate the compressibility at the volumes listed in the tables below. *Compare this to the compressibility of the van der Waals gas and calculate the percentage difference between the two values.*

Volume (L)	Z N₂	Z vdW	% diff.
4.000			
3.000			
2.000			
1.000			
0.880			

Calculate the following values at 500 K.

Volume (L)	Z N₂	Z vdW	% diff.
4.000			
2.000			
1.000			
0.100			
0.049			

Calculate the following values at 1200 K.

Volume (L)	Z N₂	Z vdW	% diff.
4.000			
2.000			
1.000			
0.500			
0.081			

4. *What is the general trend of the difference between the compressibility of the real gas and the compressibility of the van der Waals gas as the temperature increases? Why?*

_____

_____

_____

_____

5. *What is the general trend as the volume decreases? Why?* _____

_____

_____

_____

6. *Under what conditions is the van der Waals equation the best approximation for the behavior of the real gas N₂?* _____

7. At 100 K, 500 K, and 1200 K, the volumes 0.880 L, 0.049 L, and 0.081 L, respectively, are within 1 mL of the minimum volume of N_2 at each respective temperature. *What causes the wide disparity between the percent differences at 0.880 L, 0.049 L, and 0.081 L? Why are these differences significantly different from those at the other volumes?*

_____

_____

_____

_____

_____

_____

_____

Thomson

As scientists began to examine atoms, their first discovery was that they could extract negatively charged particles from atoms. They called these particles electrons. In order to understand the nature of these particles, they wanted to know how much they weighed and how much charge they carried. Thomson showed that if you could measure how much a beam of electrons were bent in an electric and magnetic field, you could figure out the ratio of mass to charge for the particles. You will repeat some of Thomson's experiments in this lab.

1. Set up the optics table for the Thomson experiment by placing the electron gun on the table, aimed at the phosphor screen, and placing the electric and magnetic fields between them right in front of the phosphor screen.

2. Turn on the phosphor screen, and push the grid button.

3. Set the electron gun energy to 100 eV with an intensity of 1 nA.

4. Increase the voltage of the electric field to 10 V.

   What happens to the spot from the electron gun? _____

   How is the electric field calculated from the applied voltage? _____

   Which direction is the electric field pointing? _____

   What is the force produced by an electric field? _____

   What voltage do you have to apply to move the spot to the first line in the grid? _____

   What voltage is necessary to move it just off the screen? _____

5. Increase the electron gun energy to 500 eV.

   How does increasing the electron gun energy change the speed of the electrons? _____

   _____

   How does this increase change the deflection of the electrons? _____

   Why does the deflection change? _____

   _____

   What voltage do you need to deflect the electrons to the edge of the screen? _____

   To the first grid line? _____

6. Decrease the electron gun energy to 10 eV.

   How does this change the deflection of the electrons? _____

7. Choose at least five other electron energies, and find the voltages necessary to deflect the electron beam to the edge of the screen. Then plot a graph of electron energy versus voltage.

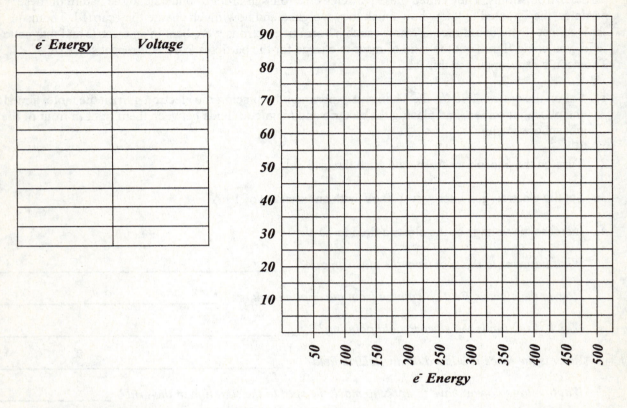

e⁻ Energy	Voltage

What is the trend of the data? _____

8. Using your graph, predict the voltage necessary to deflect a 235 eV electron beam to the edge of the screen, and then test to determine whether your prediction is right. Finally, predict the electron energy necessary to have the beam deflected to the edge of the screen by a voltage of 20 V, and test to determine whether your prediction is right.

(235 eV electron beam) Voltage prediction: _____ *Measured voltage:* _____

(20 V) Electron energy prediction: _____ *Measured electron energy:* _____

9. Turn off the electric field, and repeat the experiment with the magnetic field. Set the electron gun energy back to 100 eV, and turn the magnetic field on to 20 μT.

What happens to the spot from the electron gun? _____

Which direction is the magnetic field pointing? _____

What is the force produced by a magnetic field? _____

10. As before, choose several electron energies, and find the magnetic fields necessary to deflect the electron beam to the edge of the screen. Then, plot a graph of electron energy versus field.

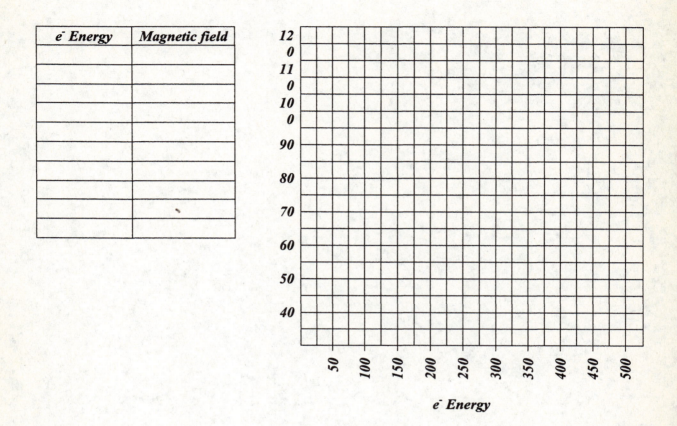

e⁻ Energy	Magnetic field

How does the magnetic field change the electron beam differently than the electric field? _____

_____

Rutherford Backscattering

A key experiment in understanding the nature of atomic structure was completed by Ernest Rutherford in 1911. He set up an experiment that directed a beam of alpha particles (helium nuclei) through a gold foil and then onto a detector screen. He observed that alpha particles were not only emerging in the direction that he expected, but that he could detect alpha particles at all angles, even straight backwards. He described this as ". . . almost as incredible as if you fired a 15-inch shell at a piece of tissue paper and it came back and hit you." He suggested that the experiment could be understood if almost all of the mass of an atom was concentrated in a small, positively charged central nucleus. In this experiment, you will make observations similar to those of Professor Rutherford.

1. Set up the Rutherford experiment by placing the alpha-particle source on the optics table, pointed at the foil holder with a gold foil, and placing the phosphor screen behind the gold foil to detect the alpha particles coming through the foil.

2. Turn on both the alpha source and the phosphor screen. Observe the screen with the alpha-particle beam shining directly through the foil and into the phosphor screen.

   What do the different signals on the screen mean? _____

   _____

3. Now change the detector to a different location. (If you don't see a signal for a position, you might have to turn on the persist button, and wait for a few minutes.)

   What differences do you see in the signal? _____

   _____

   How many distinct locations can you put the detector in? _____

   How does the signal depend on the angle formed by the source/foil/detector combination? _____

   _____

4. Using the persist button, it is possible to estimate the number of particles that hit the screen as a function of angle. You can do this by counting the particles that hit the screen over a given length of time. Now graph this rate (in particle hits per second) as a function of angle.

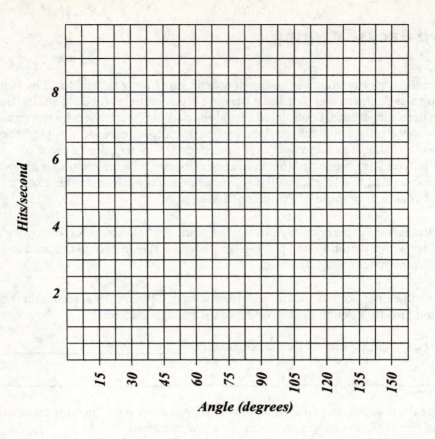

Hits/second

Angle (degrees)

What percentage of the particles is being scattered backwards?

---

5. Assuming that scattering backwards means that the alpha particle hits a nuclei head on, and that the metal foil is 0.001 cm thick, estimate the diameter of a gold nucleus.

Note: The atomic diameter of a gold atom is about 2.88×10^{-10} m.
Calculations:

$$layers = \frac{0.00001}{2.88 \, x \, 10^{-10}} \, ;$$

$$\frac{A_{nucleus}}{A_{atom}} = \frac{(\# \, scattered \, / \, \sec) \, / \, layers}{100,000} \, ;$$

$$d_{nucleus} = d_{atom} \sqrt{\frac{\# \, scattered \, / \, \sec}{100,000 \, x \, layers}} \, .$$

Gold nucleus diameter: _____

6. If R is the radius of an atomic nucleus, r_0 is the radius of a nucleon, and A is the atomic number, show how R can be approximated by $R = r_0 A^{1/3}$.

7. Given that the value of r_0 is approximately 1.4 fm, calculate the size of the gold nucleus.

    Expected size of gold nucleus: _____

    How well does your previous calculation from your measurements agree with this expected value?

    _____

    Why is your measured value not the same as the expected value? _____

    _____

Photoelectric Effect – 1

Though Einstein is most famous for his work in describing relativity in mechanics, his Nobel Prize was for understanding a very simple experiment. It was long understood that if you directed light of certain wavelength at a piece of metal, it would emit electrons. Several inconsistencies in the results were known, which led Einstein to suggest that we need to think of light as being composed of particles and not just as waves. You will have a chance to recreate some of the measurements that led to Einstein's theory.

1. Set up the photoelectric effect experiment by placing the laser, a Sodium (Na) foil, and the phosphor screen detector on the table. You need to place them so that the laser and the phosphor screen are each at about a 45-degree angle to the foil. Turn on the phosphor screen. Set the laser power to 1 nW and set the laser to the largest wavelength which still gives a signal.

    What wavelength (to within 1 nm) is the largest that still gives a signal? _____

    What is the equation that relates the wavelength of a photon to the energy of a photon? _____

    We define the work function to be the minimum energy necessary to remove an electron from the metal.

    In nm what is the work function of Na? _____

    What is it in eV? _____

    Based on what you know about atoms, which would you predict has the smallest work function, Na,

    K, Rb, or Cs? _____

2. Measure the work functions for Na, K, Rb and Cs. Record the values in eV units.

    Na _____ *K* _____ *Rb* _____ *Cs* _____

    Do these values agree with your chemical intuition? _____

    _____

3. *Predict the order of increasing work functions for Co, Ni, Cu and Zn, and then measure the work functions for these elements. Again, record the values in eV units.*

    Ranking prediction: _____

    Measured values for the work functions:

    Co _____ *Ni* _____ *Cu* _____ *Zn* _____

4. Make a graph of work function versus atomic number for every available metal foil.

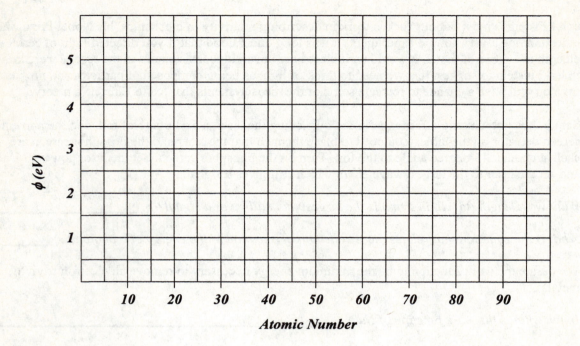

Describe the periodic trends in the results. _____

_____

_____

Can you make any conclusions or generalizations about the trends you observed? _____

_____

Photoelectric Effect – 2

Though Einstein is most famous for his work in describing relativity in mechanics, his Nobel Prize was for understanding a very simple experiment. It was long understood that if you directed light of certain wavelength at a piece of metal, it would emit electrons. Several inconsistencies in the results were known, which led Einstein to suggest that we need to think of light as being composed of particles and not just as waves. You will have a chance to recreate some of the measurements that led to Einstein's theory.

1. Set up the photoelectric effect experiment by placing the laser, a Sodium (Na) foil, and the bolometer on the table. The laser and the bolometer need to each be at a 45-degree angle to the face of the metal foil. Turn on the laser, and set the power to 1 nW and the wavelength to 400 nm.

2. The bolometer measures the kinetic energy of electrons emitted from the metal. You should see a peak on the bolometer detector screen. Zoom in on the area of the peak so that you can accurately read the kinetic energy of the electrons. Record this value.

    Kinetic energy of electrons: _____

3. Einstein suggested that the energy of the emitted electrons was the energy of a photon of light minus the work function of the metal, or the energy that binds the electrons to the metal. Calculate the work function (in units of eV) by taking the difference of the energy of a photon from the laser minus the kinetic energy of an electron.

    Work function for Na: _____

4. Measure the electron kinetic energy at five different wavelengths of light (less than 450 nm), and calculate the work function (ϕ) in units of eV.

λ	100	200	250	300	350
Energy of photon					
Kinetic energy of electron					
ϕ					

Is the work function independent of wavelength? _____

Based on what you know about atoms, which would you predict would have the smallest work

function, Na, K, Rb or Cs? _____

5. Measure the work functions for Na, K, Rb and Cs.

    Na _____ *K* _____ *Rb* _____ *Cs* _____

Do these values agree with your chemical intuition? _____

_____

6. Predict the order of increasing work functions for Co, Ni, Cu and Zn, and then measure the work functions for these elements.

   Ranking prediction: _____

   Measured values for the work functions:

   Co _____ *Ni* _____ *Cu* _____ *Zn* _____

7. Make a graph of work function versus atomic number for every available metal foil.

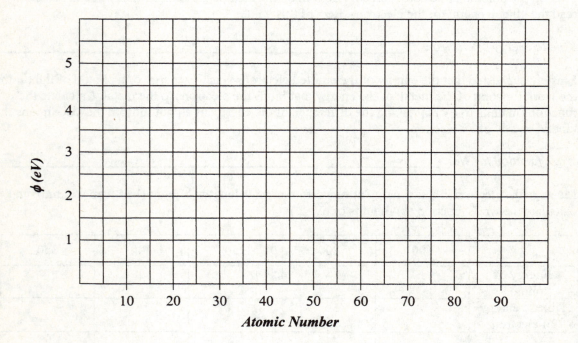

Describe the periodic trends in the results. _____

_____

_____

Can you make any conclusions or generalizations about the trends you observed? _____

_____

de Broglie – 1

de Broglie was the first person to suggest that particles could be considered as having wave properties. Specifically, he suggested that $\lambda = constant / p$ (wavelength is inversely proportional to momentum). In this assignment, you will calculate the constant that relates λ to p.

1. Set up the optics table to measure the diffraction of electrons with an energy of 2 meV using the electron gun, the two-slit device, and the phosphor screen. Set the current (intensity) of the electron gun to at least 1 nA.

   What is the kinetic energy of the electrons in Joules? _____

2. Adjust the slit spacing, and observe how the diffraction pattern changes accordingly.

   How does the diffraction pattern change when you increase the spacing between the slits? _____

   _____

3. Find a slit spacing that gives 7 to 11 well-defined diffraction fringes.

   Draw a picture of the diffraction pattern.

   Slit spacing: _____

   What characteristic of the electron accounts for the diffraction pattern created by the two-slit

   experiment? _____

4. Now, change the electron gun for the laser, and the phosphor screen for the camera. Set the intensity of the laser to at least 1 nW.

5. Keeping the slit spacing the same, find the wavelength of light that gives the same diffraction pattern.

   Wavelength: _____

   How is this wavelength related to the wavelength of the electrons? _____

   _____

6. Given that $E_{kinetic} = p^2/2m$, solve for p and then calculate the constant that relates p with λ.

Calculations:

Constant = _____

de Broglie – 2

de Broglie was the first person to suggest that particles could be considered as having wave properties. Specifically, he suggested that $\lambda = constant / p$ (wavelength is inversely proportional to momentum). In this assignment, you will calculate the constant that relates λ to p.

1. Set up the optics table to measure the diffraction of electrons using the electron gun, the two-slit device, and the phosphor screen. Set the current of the electron gun to at least 1 nA. Set the slit spacing to 100 nm.

2. Adjust the kinetic energy of the electrons.

   What happens to the diffraction pattern as you increase the energy of the electrons? _____

   _____

   Does this support de Broglie's equation? If so, how? _____

   _____

   _____

   Express the constant in de Broglie's equation as a function of kinetic energy, mass, and wavelength.

3. Set the electron gun energy between 1 and 4 meV. Find a slit spacing that gives 7 to 11 well-defined diffraction fringes. Then, by using the laser and video camera, find the wavelength that gives this same diffraction pattern for this particular slit spacing. Repeat this procedure for two different kinetic energies, and record the values in the following table. Then calculate the constant that relates p with λ.

Electron Kinetic Energy	Slit Spacing	Diffraction Pattern	Wavelength	Constant

   Averaged value for constant: _____

4. This constant that you have calculated is known as Planck's constant. Look up its actual value and compare it with your value.

   % deviation from Planck's constant: _____

HCl Gas Absorbance

HCl gas does not absorb visible light, but it does absorb infrared light. When it absorbs one photon of infrared light to go from the ground vibrational state to the first excited vibrational state, it can also change rotational states. These rotations are also quantized, meaning that molecules can only rotate at certain frequencies. In this lab, you will measure the rotational energy changes that accompany vibrational changes.

1. Go to the stockroom, and check out the super light bulb, the gas cell with HCl gas, and the spectrometer. Set up the experiment with the light shining through the gas and into the spectrometer. Turn on the spectrometer, and set the units to frequency.

    Draw a picture of what you see.

There are two large sets of peaks due to the hydrogen atoms. Hydrogen has two different isotopes (different nuclear weights). We label these H (for hydrogen), which has a mass of 1 amu (atomic mass unit), and D (for deuterium), which has a mass of 2 amu.

Which one would you expect to absorb at the lower frequency and why? _____

_____

_____

_____

Which isotope is more abundant? _____

Do the relative intensities of the two sets of peaks correlate to their natural abundances? Why or why

not? _____

_____

2. Zoom in on the lower frequency peaks.

   Draw a picture of what you see.

   Chlorine atoms also come in two isotopes with masses of 35 and 37 amu. By zooming in sufficiently, you will notice that each main peak is a doublet.

   Predict which peaks of the doublets belong to ^{35}Cl and which to ^{37}Cl. _____

   _____

   Which is more abundant? _____

   Does this agree with the mass for chlorine that you see on a periodic table (which is an average mass

   based on natural abundances)? _____

   _____

   _____

   Both large sets of peaks have two branches of peaks. The branch of lower frequency peaks have a change from a higher rotational state in the ground vibrational state to a lower one in the first excited vibrational state, and the higher frequency peaks have a change from a lower rotational state in the ground vibrational state to a higher one in the first excited vibrational.

   Which branch of peaks is more intense? _____

   Why? _____

   _____

   _____

   _____

I₂ Gas Absorbance

I_2 gas is interesting because it absorbs visible light, which causes an electron to move between two electronic energy levels. At the same time, the molecule can change vibrational energy. Because the vibrations are quantized (the molecule can only vibrate at certain frequencies), the spectrum is not continuous, but instead has peaks where the vibrations change from one energy level to another.

1. Go to the stockroom and bring out the super light bulb, the gas cell with I_2 gas, and the spectrometer. Set up the experiment with the light shining through the gas and into the spectrometer.

2. Carefully zoom in on the spectrum.

    Draw the basic structure that you see.

3. Increase the power of the super light bulb.

    Does the spectrum change? Why or why not? _____

    _____

    _____

4. Change into frequency units, and zoom in on the area of the spectrum near where it absorbs the most light (where the transmittance is the lowest). You should see a series of peaks. Using the cursor and the *x-y* scale, measure the difference in frequency between neighboring peaks for the first 10 well-resolved peaks (highest energy peaks). Numbering the peaks from 1 to 10, plot the differences in frequency against peak number.

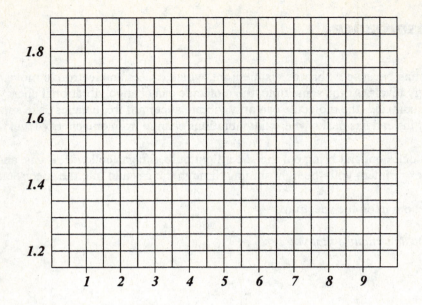

What does the function look like? _____

What happens to the difference in frequency between neighboring peaks at higher frequency peaks?

_____

What happens to the energy difference between neighboring peaks at higher frequency peaks? _____

_____

What does this tell you about vibrational modes within an electronic state? _____

_____

_____

Water Absorption

Water absorbs light in the infrared region of the electromagnetic spectrum. Absorption spectroscopy in the infrared region is called infrared (IR) spectroscopy. Infrared radiation causes the bonds within molecules to vibrate. For this reason, IR spectroscopy is sometimes called vibrational spectroscopy. The atoms within molecules are always moving. The bonds between atoms will absorb light if the light is at the same frequency as the frequency of the vibration of the bond. Thus, bonds act like springs, and just the right amount of energy must be added to make them vibrate. The absorption of this energy makes the vibration have greater amplitude, yet the vibration remains at the same frequency and wavelength. A certain functional group will always absorb within the same general region of the spectrum. For example, the O—H group gives a strong absorption peak around 2778 nm to 3125 nm.

1. Set up the optics table for this experiment by selecting *Absorption in Liquids – Water* on the clipboard of preset experiments.

    What source is used in this experiment? _____

    How is the light produced by this source different than the light produced by the laser? _____

    _____

    _____

    Which liquid sample is being used in this experiment? _____

    What detector is used in this experiment and what does it measure? _____

    _____

    Draw a rough sketch of the spectrum.

    Between what wavelengths is the first wide absorption peak? _____

    What causes this absorption peak? _____

2. On the spectrometer screen, switch the toggle from *FULL* to *VISIBLE*.

    What do you observe? _____

Would you expect water to absorb light in the visible region? Why or why not? _____

_____

Raman Scattering

The vibrational modes of a molecule are quantized, which means that molecules can only vibrate at certain frequencies. In normal light absorbance spectroscopy, absorbance peaks are observed at frequencies of light that have the correct amount of energy to make molecules vibrate one step faster. C. V. Raman, a scientist from India, was the first person to demonstrate another type of spectroscopy, and so it is named after him. In this measurement, you send laser light in the visible region through a sample. Most of the photons in the beam travel through the sample, but a small number interact with the sample and are scattered in different directions. An even smaller number (less than one in a million) interact with the sample and during the process either absorb one vibrational energy quantum from a molecule (leading to photons with a bit more energy) or give one quantum of energy to a molecule (leading to photons with a bit less energy). In the Raman spectra, make sure to zoom in to the spectra in the wings around the central peak to see the other peaks. These satellite peaks are very small.

1. Go to the stockroom; bring out the laser, the gas cell filled with HCl, and the spectrometer; and arrange them on the table with the laser directed into the gas cell, and the laser and spectrometer at a 90-degree angle from each other.

2. Set the laser power to 1 nW and the wavelength to 620 nm. Turn on the spectrometer.

   Draw a picture of what you see.

   What causes each peak? _____

   _____

   _____

   _____

3. Increase the power of the laser.

   Does this change the spectrum? Why or why not? _____

   _____

4. Now change the wavelength of the laser.

    Does this change the spectrum? If so, how? _____

    _____

5. With the spectrometer set for wavelength, measure the difference between the main peak and the satellite peaks using laser wavelengths of 620, 570, 520, 470 and 420 nm. Only measure the satellite peak to the left.

Laser Wavelength	620	570	520	470	420
Difference (in λ)					

    Is the difference the same at different laser wavelengths? _____

6. Make a plot of wavelength versus the difference.

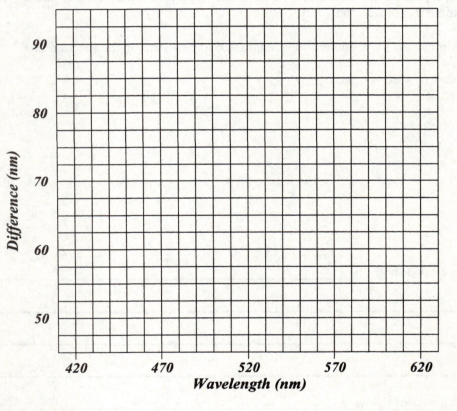

    Is the function linear? _____

    How is wavelength related to energy? _____

7. With the spectrometer set for frequency, measure the difference between the main peak and the satellite peaks in terahertz (THz) using the same laser wavelengths as before (620, 570, 520, 470, and 420 nm). Again, only measure the satellite peak to the left.

Laser Wavelength	620	570	520	470	420
Difference (in ν)					

Is this difference constant? Why should this be expected? _____

_____

What is the difference in the Raman spectrum in THz? _____

What wavelength of light has this frequency? _____

8. Exchange the laser with the super light bulb, and measure the absorbance spectrum by placing the spectrometer in line with the super light bulb and the sample.

   Do you see an absorbance peak at the wavelength that you specified earlier? _____

   Why or why not? _____

   _____

   _____

   _____

9. Switch the super light bulb for the laser again. Set the wavelength of the laser 100 nm below that of your specified wavelength. Set the intensity of the laser to at least 1 nW. Then exchange the spectrometer for the photodiode. Turn on the photodiode, and keep it in line with the laser and sample.

   What is the function of the photodiode? _____

   _____

10. Slowly increase the wavelength of the laser 200 nm.

    What do you observe? _____

    What causes the differences in intensity? _____

    _____

    _____

Now we will do some measurements with liquids.

11. Return the gas cell and photodiode to the stockroom, and replace it with the liquid cell and spectrometer, respectively. Fill the cell with Benzene (C_6H_6), and return both to the laboratory. Set up the experiment again for a Raman experiment using a laser wavelength of 620 nm.

Draw a picture of the spectrum.

12. Measure the difference in frequency between each satellite peak in the spectrum and the main peak (the frequency of the laser beam).

Main peak frequency: _____

Satellite peak number	ν	ν of symmetrical peak	Difference between main and satellite ν's
1			
2			
3			
4			
5			
6			
7			
8			
9			
10			
11			

13. Look at the spectrum of each of the liquids, and count the number of satellite peaks for each one.

Which has the most peaks? _____

Which has the least? _____

What are the differences between liquids and gases in a Raman spectrum?

_____

Why? _____

	# of satellite peaks
C_6H_6	
H_2O	
CCl_4	
C_6H_{12}	
THF	
MeOH	
CH_3CN	
C_6H_{10}	